Current Topics in Microbiology and Immunology

178

Editors

R. W. Compans, Birmingham/Alabama · M. Cooper,
Birmingham/Alabama · H. Koprowski, Philadelphia
I. McConnell, Edinburgh · F. Melchers, Basel
V. Nussenzweig, New York · M. Oldstone,
La Jolla/California · S. Olsnes, Oslo · M. Potter,
Bethesda/Maryland · H. Saedler, Cologne · P. K. Vogt,
Los Angeles · H. Wagner, Munich · I. Wilson,
La Jolla/California

Cover illustration:
Scanning EM of complement coated erythrocytes that have formed a rosette with a neutrophil. Adherence was mediated by receptors on the neutrophil that recognized proteins that were deposited onto the erythrocytes as a consequence of complement activation. The presence of activation and degradation products of complement C3 on target cells bearing IgG antibodies greatly enhances immune adherence and ingestion by phagocytes.

Charles J. PARKER, A. B., M.D.
Associate Professor of Medicine
University of Utah School of Medicine
Chief, Hematology Subsection
Department of Veterans Affairs Medical Center
Salt Lake City, Utah 84148
USA

ISBN 3-540-54653-7 Springer-Verlag Berlin Heidelberg New York
ISBN 0-387-54653-7 Springer-Verlag New York Berlin Heidelberg

This work is subject to copyright. All rights are reserved, whether the whole or part of the material is concerned, specifically the rights of translation, reprinting, reuse of illustrations, recitation, broadcasting, reproduction on microfilm or in any other way, and storage in data banks. Duplication of this publication or parts thereof is permitted only under the provisions of the German Copyright Law of September 9, 1965, in its current version, and permission for use must always be obtained from Springer-Verlag. Violations are liable for prosecution under the German Copyright Law.

© Springer Verlag Berlin Heidelberg 1992
Library of Congress Catalog Card Number 15-12910
Printed in Germany

The use of general descriptive names, registered names, trademarks, etc. in this publication does not imply, even in the absence of a specific statement, that such names are exempt from the relevant protective laws and regulations and therefore free for general use.

Product liability: The publishers cannot guarantee the accuracy of any information about dosage and application contained in this book. In every individual case the user must check such information by consulting the relevant literature.

Typesetting: Thomson Press (India) Ltd, New Delhi; Offsetprinting: Saladruck, Berlin; Bookbinding: Lüderitz & Bauer, Berlin.
23/3020-5 4 3 2 1 0 – Printed on acid free paper.

Membrane Defenses Against Attack by Complement and Perforins

Edited by C. J. Parker

With 26 Figures

Springer-Verlag
Berlin Heidelberg New York
London Paris Tokyo
Hong Kong Barcelona
Budapest

List of Contents

C. J. PARKER: Regulation of Complement
by Membrane Proteins: An Overview 1

A. NICHOLSON-WELLER:
Decay Accelerating Factor (CD55) 7

G. D. ROSS: Complement Receptor Type 1 31

M. K. LISZEWSKI and J. P. ATKINSON:
Membrane Cofactor Protein 45

M. H. HOLGUIN and C. J. PARKER: Membrane Inhibitor
of Reactive Lysis 61

L. S. ZALMAN: Homologous Restriction Factor 87

D. V. DEVINE: The Effects of Complement Activation
on Platelets 101

B. P. MORGAN: Effects of the Membrane Attack Complex
of Complement on Nucleated Cells 115

D. M. LUBLIN: Glycosyl-Phosphatidylinositol Anchoring
of Membrane Proteins 141

W. F. ROSSE: Paroxysmal Nocturnal Hemoglobinuria . . 163

E. R. PODACK: Perforin-Structure, Function,
and Regulation 175

Subject Index 185

List of Contributors

(Their addresses can be found at the beginning of their respective chapters)

ATKINSON, J. P.	45	NICHOLSON-WELLER, A.	7
DEVINE, D. V.	101	PARKER, C. J.	1, 61
HOLGUIN, M. H.	61	PODACK, E. R.	175
LISZEWSKI, M. K.	45	ROSS, G. D.	31
LUBLIN, D. M.	141	ROSSE, W. F.	163
MORGAN, B. P.	115	ZALMAN, L. S.	87

Regulation of Complement by Membrane Proteins: An Overview

C. J. PARKER

The complement system has evolved as the major humoral defense mechanism against infection, and complement activation products mediate many of the processes of inflammation (MÜLLER-EBERHARD 1988). For example, C3a and C5a are potent anaphylatoxins, C5a is the primary mediator of neutrophil chemotaxis, and activation (C3b) and degradation (iC3b) fragments of C3 are critically important opsonins that are recognized by specific receptors on phagocytic cells (ROSS and MEDOF 1985). The complement system can also mediate cell killing directly through formation of the membrane attack complex.

Because of its enormous destructive capacity, elaborate safeguards for confining complement activity to the desired target have evolved. Complement activation is mediated by either the classical or the alternative pathway, but the conditions required for activation of the two systems are very different. Activation of the classical pathway of complement (CPC) is highly specific in that antibody is required to initiate the process. Thus, the system is quiescent unless there is invasion by a foreign organism against which the host has been immunized. Accordingly, under normal physiological conditions, host cells do not require protection against the CPC since antibodies against self do not arise.

In contrast to the CPC, there is no initiating factor for the alternative pathway of complement (APC); consequently, the APC does not actively discriminate between host and foreign cells. Rather, the burden for avoiding attack by the APC is borne directly by cells that must express membrane properties to restrict the activity of the system.

In plasma, the APC is in a state of continuous, low-grade activation; thus, the system is primed for attack at all times. The activity of the system is kept at a low level because, in the fluid phase, the formation and stability of the amplification C3 convertase is tightly controlled by the endogenous plasma regulatory proteins factor H and factor I. Nonetheless, some low-grade activation does occur because an internal thioester in native C3 is subject to spontaneous hydrolysis (Fig. 1). As a consequence, the nascent (H_2O) C3 molecule undergoes a conformational change that initially exposes a magnesium-dependent binding site for factor B. The (H_2O) C3·B complex is subject to rapid inactivation because

---

Hematology/Oncology Section (111c), VA Medical Center, 500 Foothill Drive, Salt Lake City, Utah 84148, USA

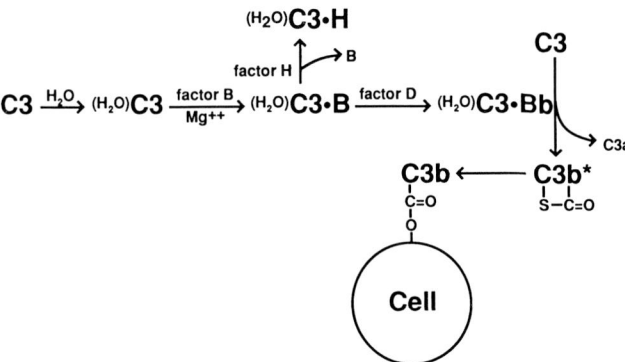

Fig. 1. Proposed mechanism of spontaneous activation of the alternative pathway of complement (APC). The spontaneous hydrolysis of plasma C3 induces a conformational change in the nascent (H_2O)C3 molecule such that a binding site for factor B becomes exposed. The amplification C3 convertase of the APC, (H_2O)C3b·Bb, is formed when factor B is activated by factor D. The activity of the convertase is very limited due both to its intrinsic instability and to the regulatory effects of factor H and factor I. Nonetheless, this process appears to be sufficient to perpetuate continuous, albeit low-grade, activation of C3. Nascent C3b that is generated by the fluid phase C3 convertase can form a covalent bond with a cell surface acceptor molecule. This binding appears to be random and indiscriminate. Host cells are protected from injury mediated by the APC because they have evolved surface proteins that inhibit the activity of C3 convertase and the membrane attack complex

hydrolysis of C3 also exposes a binding site for factor H, and binding of factor H to (H_2O) C3 displaces B from the complex. Binding to (H_2O) C3 causes factor B to undergo a conformational change that exposes a cleavage site for factor D. The catalytic subunit (Bb) of the APC convertase is subsequently generated by the enzymatic proteolysis of factor B by factor D (factor D appears to exist only in an active state since a zymogen form has not been conclusively identified). Although the half-life of the complex is brief (owing to the intrinsic instability of the complex and to the regulatory effects of factor H), apparently enough (H_2O) C3·Bb is formed to mediate continuous low-grade activation of C3.

The activation process is initiated when native C3 binds to the (H_2O)C3·Bb complex (Fig. 1). The subsequent enzymatic cleavage of C3a from native C3 induces a conformational change in the resulting C3b molecule such that the internal thioester becomes exposed. This exposed but intact thioester constitutes the labile binding site of nascent C3b. The binding site has a half-life measured in milliseconds and usually becomes inactive as a result of hydrolysis. By acyl transfer, however, the thioester also has the capacity to mediate covalent binding to a nearby cell surface through formation of an ester or imidoester bond. Under physiological conditions, formation of ester bonds is favored (thus accounting for the propensity of C3b to bind to cell surface carbohydrates), but binding is otherwise nonspecific. Thus, nascent C3b can bind equally well to both host and foreign cells. Inasmuch as C3b is the nidus for formation of the amplification C3 convertase of the APC, once C3b has bound to the cell surface, the system will be activated and amplified unless membrane and plasma factors work in concert to control it.

The virulence of invasive organisms depends largely upon their capacity to circumvent attack by complement, and successful pathogens have devised a number of ingenious strategies for evading complement-mediated destruction (JOINER 1988).

The importance of the complement system is further emphasized by the evolutionary pressure that was placed on humans to develop mechanisms for protecting self against complement-mediated injury. Accordingly, human cells (particularly those that are present at sites of inflammation, e.g., hematopoietic and endothelial cells) have evolved highly specialized membrane constituents that act independently or in concert with plasma regulatory proteins to inhibit the functional activity of complement. Decay accelerating factor (DAF, CD55) and complement receptor type 1 (CR1, CD35) directly inhibit the formation and stability of the convertase, while CR1 and membrane cofactor protein (MCP, CD46) act indirectly by serving as cofactors for the enzymatic degradation of C3b to iC3b by factor I. (DAF, CR1, and MCP are reviewed in this volume.)

In an analogous fashion, DAF, CR1, and MCP control the formation and stability of the amplification C3 convertase of the CPC because they can bind to both C4b and C3b. (CR1 and MCP also act as cofactors for enzymatic degradation of C4b by factor I.) Further, inasmuch as C3b is also a common constituent of the C5 convertases of both the APC and the CPC, the membrane regulators DAF, CR1, and MCP also inhibit C5 activation (Fig. 2). As discussed above, protection of host cells against attack by the CPC is usually not necessary since antibodies against self are normally not present. However, under certain pathological circumstances (e.g., autoimmune hemolytic anemia) membrane regulators of the CPC become critically important by virtue of their capacity to limit the severity of complement-mediated injury.

Failure of a cell to regulate the amplification C3 and C5 convertases allows the potentially cytolytic membrane attack complex (MAC) to be generated on its surface (BHAKDI and TRANUM-JENSEN 1983). The complex is formed as a result of the spontaneous and sequential binding of C5b, C6, C7, C8 (in 1:1:1:1 stoichiometry), and multiple molecules of C9 (the number per complex appears to be variable and estimates range from 3 to 16). The multiple molecules of C9 undergo polymerization, and this process forms the transmembrane channel that appears as the classical "doughnut" lesion visualized by electron microscopy. Two membrane proteins, homologous restriction factor (HRF), also identified as C8 binding protein (C8bp) and MAC inhibitory protein (MIP), and membrane inhibitor of reactive lysis (MIRL, CD59), have the capacity to regulate the assembly of the MAC, at least in part by inhibiting the binding and polymerization of C9. (HRF/C8bp and MIRL are reviewed in this volume.)

DAF, CR1, and MCP share a significant amount of primary sequence homology. In addition, they have a strikingly similar motif structure, and it is this feature that confers the capacity to bind C3b (and C4b). The genes for these proteins (along with several others that encode complement regulatory proteins

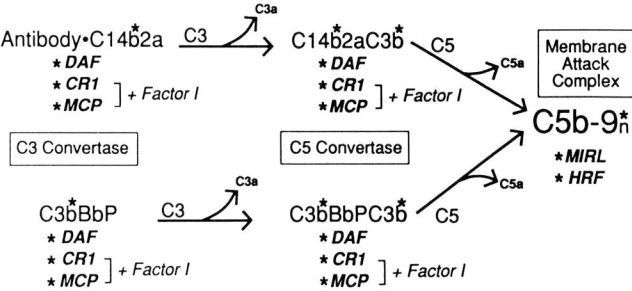

Fig. 2. Regulation of the classical (CPC) and alternative (APC) pathways of complement by membrane proteins. Activation of the CPC is highly specific in that antibody is required to initiate the process. Therefore, membrane regulatory proteins on host cells are not normally required to control the CPC since antibodies against self do not arise except under pathological conditions (e.g., autoimmune hemolytic anemia). In contrast, the APC does not discriminate between host and foreign cells. Rather, the burden for avoiding attack by the APC is born directly by the cells. Consequently, membrane proteins that regulate the APC C3 and C5 convertases and the membrane attack complex have evolved. These same proteins regulate the CPC because the two pathways share components that are identical (e.g., C3) or highly homologous both structurally and functionally (e.g., C4 compared to C3 and C2 compared to factor B). The *asterik* denotes the component within the complex with which the regulatory protein interacts. MCP and CR1 serve as cofactors for the factor I-mediated degradation of C4b and C3b

with similar motif structures) are clustered on the long arm of chromosome 1, suggesting that they evolved by gene duplication from a common ancestor (HOURCADE et al. 1989).

The derived amino acid sequence of MIRL, but not HRF, has been reported. Thus, it is unknown if MIRL and HRF share primary sequence homology and motif structure in a manner analogous of DAF, CR1, and MCP. MIRL does not share sequence homology with any other complement proteins; however, a modest degree of homology (approximately 25%) exists between MIRL and a multigenic family of murine lymphocyte proteins called LY-6.

The importance of membrane regulators of complement is manifested by the pathophysiological consequences associated with paroxysmal nocturnal hemoglobinuria (PNH), a disease in which DAF, HRF, and MIRL are deficient. Since they lack the membrane constituents that control the activity of the C3 and C5 convertases and the MAC, PNH erythrocytes undergo spontaneous complement-mediated lysis. (The life span of PNH erythrocytes is 4–6 days compared to 120 days for normal erythrocytes.) Consequently, the primary clinical manifestation of PNH is chronic intravascular hemolytic anemia that, in most cases, produces a significant amount of morbidity. (PNH is reviewed in this volume.)

Over the last decade a number of provocative observations have been made that make a review of membrane regulators of complement particularly timely. First, the primary sequence of all of the membrane regulators except HRF has been derived, and, as discussed above, CR1, DAF, and MCP belong to the gene superfamily known as the regulators of complement activation (RCA), located on the q32 band of chromosome 1. Other members of the family include factor B, C2, factor H, and C4 binding protein, and all of these proteins have in common a motif structure that confers upon them the capacity to interact with C3, C4, or both.

A second recent watershed event was the finding that DAF is anchored to cell membranes through a glycosyl phosphatidylinositol (GPI) moiety (reviewed in this volume). This observation was important not only because it provided information about the structure of DAF, but also because it provided new insight into the molecular basis of PNH. The circumstances that led to this discovery serve as an example of how the identification and characterization of membrane complement regulatory proteins have been closely intertwined with the study of PNH. It had been know since 1959 that PNH erythrocytes were deficient in acetylcholinesterase (AChE) activity (AUDITORE and HARTMANN 1959). The finding, in 1983, that PNH erythrocytes were also missing DAF suggested that the defect in PNH might involve an abnormality of a structural element common to both molecules. In 1984, MEDOF and colleagues reported that isolated DAF spontaneously reincorporated into erythrocyte membranes and subsequently expressed its inhibitory activity. At about the same time as that report, AChE was also shown to spontaneously reincorporate into cell membranes. Earlier, AChE had been shown to be a member of a newly described family of amphipathic membrane proteins that shared the common structural feature of being anchored to the membrane through a GPI moiety, and the capacity of AChE to reincorporate into cell membranes was attributed to the presence of the lipid anchoring mechanism. The link between DAF and AChE was made in 1986 by DAVITZ et al. and MEDOF et al. when those investigators presented evidence that DAF was also a GPI-anchored protein. The results of those insightful studies suggested a plausible explanation for the PNH defect. According to the new paradigm, PNH arises as a result of a somatic mutation involving the hematopoietic stem cell. The mutation renders absent or dysfunctional an element that is critical in the pathway of biosynthesis or attachment of the GPI anchor to the protein backbone. To date, this paradigm has remained viable, because all proteins deficient from PNH cells are GPI-linked and, conversely, all GPI-linked hematopoietic elements have been shown to be deficient in PNH (although the deficiency is more commonly partial rather than absolute). The precise biological significance of the GPI anchor is the subject of intense investigation. That membrane constituents with complement regulatory activity are GPI-linked, however, suggests that this structural feature may be advantageous for this group of proteins which must patrol the cell and restrict the activity of complement over the entire surface of the membrane.

Over the last several years it has become evident that, when complement is activated on cells that are metabolically active, nonlethal injury stimulates the production of biologically active products that participate in the inflammatory process. Further, metabolically active cells have mechanisms for modulating membrane-associated complement activity that differ from those of erythrocytes. The effect of complement activation on cells that are metabolically active has been the subject of a number of provocative studies and is reviewed in this volume.

A discussion of the action and regulation of lymphocyte-mediated cytotoxicity is also particularly timely. Perforins are the effectors of cell killing by cytotoxic lymphocytes, and recent studies have demonstrated structural and functional similarities between the perforins and components of the MAC. Despite the considerable similarities, however, membrane proteins that regulate the MAC do not appear to control perforin-mediated lysis. Accordingly, a section on the regulation of cell killing by cytotoxic lymphocytes will consider the similarities and disparities between the complement and perforin systems.

A very recent development that has generated much interest is the finding that recombinant CR1 may be of therapeutic use for pathophysiological processes that are immune-mediated (WEISMAN et al. 1990). The advantage that CR1 has over its plasma analogue, factor H, its that is activity is not restricted by the carbohydrate moiety of the cell, as it appears to be with factor H. FEARON first reported the isolation of CR1 in 1979. That within 10 years a basic observation has moved into the clinical arena demonstrates both the rapid pace and the insightful thought that has characterized complement research over the last decade. The aim of this volume is to communicate the results of some of these elegant studies.

References

Auditore JV, Hartmann RC (1959) Paroxysmal nocturnal hemoglobinuria: II. Erythrocyte acetylocholinesterase defect. Am J Med 27: 401–410
Bhakdi S, Tranum-Jensen J (1983) Membrane damage by complement. Biochim Biophys Acta 737: 343–372
Davitz MA, Low MG, Nussenzweig V (1986) Release of decay-accelerating factor (DAF) from the cell membrane by phosphatidylinositol-specific phospholipase C (PIPLC): selective modification of a complement regulatory protein. J Exp Med 163: 1150–1161
Fearon DT (1979) Regulation of the amplification C3 convertase of human complement by an inhibitory protein isolated from human erythrocyte membrane. Proc Natl Acad Sci USA 76: 5867–5871
Hourcade DE, Holers VM, Atkinson JP (1989) The regulators of complement activation (RCA) gene cluster. Adv Immunol 45: 381–416
Joiner KA (1988) Complement evasion by bacteria and parasites. Annu Rev Microbiol 42: 201–230
Medof EM, Kinoshita T, Nussenzweig V (1984) Inhibition of complement activation on the surface of cells after incorporation of decay-accelerating factor (DAF) into their membranes. J Exp Med 160: 1558–1578
Medof ME, Walter EI, Roberts WL, Hass R, Rosenberry TL (1986) Decay-accelerating factor of complement is anchored to cells by a C-terminal glycolipid. Biochemistry 25: 6740–6747
Müller-Eberhard HJ (1988) Molecular organization and function of the complement system. Annu Rev Biochem 57: 321–347
Ross GD, Medof ME (1985) Membrane complement receptors specific for bound fragments of C3. Adv Immunol 37: 217–267
Weisman HF, Bartow T, Leppo MK, Marsh HC, Jr. Carson GR, Concino MF, Boyle MP, Roux KH, Weisfeldt ML, Fearon DT (1990) Soluble human complement receptor type 1: In vivo inhibitor of complement suppressing post-ischemic myocardial inflammation and necrosis. Science 249: 146–151

Decay Accelerating Factor (CD55)

A. NICHOLSON-WELLER

1 Introduction and Background	8
1.1 Complement	8
1.2 Species Restriction	8
1.3 Isolation of Membrane Inhibitors of Complement	9
2 Function of DAF	10
2.1 DAF Inhibits C3 Activation	10
2.2 Binding Ligand for DAF	12
2.3 Distinct Role of DAF in Regulating C3 Activation at the Membrane	13
3 DAF Gene	13
3.1 Localization	13
3.2 Organization	14
4 Structure of DAF	15
4.1 Predicted Peptide Structure	15
4.2 Glycan Phosphatidylinositol Anchor	15
4.3 Glycosylation of DAF	17
4.4 Alternate Forms of Mature DAF	18
5 Expression and Distribution of DAF	19
5.1 Expression on Cells	19
5.2 Regulation of Membrane Expression	19
5.3 Mobility in the Membrane	21
5.4 Can DAF Transmit Transmembrane Signals?	21
5.5 DAF in Extracellular Tissues	22
6 Abnormal Expression of DAF	22
6.1 Inab Phenotype	22
6.2 Paroxysmal Nocturnal Hemoglobinuria	23
6.3 Expression in Other Diseases	24
7 DAF in Nonhuman Species	25
7.1 Mammals	25
7.2 Parasites	25
8 Concluding Remarks	26
References	26

Beth Israel Hospital, DA-617, 330 Brookline Avenue, Boston, MA 02215, USA

1 Introduction and Background

1.1 Complement

Complement is a mediator system comprised of 12 activation proteins and at least 11 inhibitors. The activation proteins interact sequentially to generate cleavage fragments and condensation products with specific and potent biologic activity. Activated complement proteins can function as anaphylatoxins, adherence factors, chemotaxins, opsonins, and a transmembrane pore, which can be lytic (reviewed in FRIES and FRANK 1987). Regulated complement activation is essential for the survival of the host, while unregulated complement activation contributes to inflammation and is detrimental to the host. Complement is regulated by the specificity of its activation, by the lability of its activation products, and by the potent inhibitors found in the fluid phase and within the membranes of host cells. The structure, function, and expression of one important membrane regulator of complement, namely, the decay accelerating factor (DAF), will be reviewed here.

In general terms, complement is activated either by specific antibody combining with its antigen and activating C1, initiating the classical pathway, or by foreign surfaces that permit the assembly of the C3bBb enzyme, which cleaves C3 in the alternative pathway. During activation of the classical pathway, C1 cleaves C4 and the major cleavage fragment C4b binds to the target cell membrane in the vicinity of C1. C2 then binds to C4b and is cleaved by C1 to form the bimolecular enzyme C4b2a, the so-called C3 convertase, which cleaves and activates C3 and C5. C2a bears the catalytic site, and this site is active only while the C2a fragment is bound to C4b. For the alternative pathway, C3b that is spontaneously formed in the fluid phase docks on a surface which allows the subsequent binding of factor B. The latter is then activated by factor D, a circulating active serine esterase. The resultant C3bBb convertase can cleave and activate C3 and C5. Analogous to C2a of the classical convertase, it is the Bb fragment which bears the catalytic site, and that catalytic site is only active as long as the Bb fragment is bound to C3b. C3 activation is a critical step because it is where the alternative and classical pathways converge: cleavage of C3 by the classical pathway makes it potentially possible to recruit the alternative pathway, the so-called amplification loop. In addition, most of the mediators of complement are generated at the C3 cleavage step or later. The importance of regulating C3 activation is reflected in the fact that there are at least three fluid phase inhibitors and three membrane regulators, including DAF, which act at this step.

1.2 Species Restriction

It was found empirically that erythrocytes from a given species are highly resistant to complement lysis by homologous complement but susceptible to lysis by heterologous complement. This in vitro phenomenon was designated "species restriction", or "homologous restriction", and its in vivo validity is confirmed by clinicians who rarely observe intravascular hemolysis due to complement. At least two steps in the complement cascade are subject to species restriction, C3 activation (HOULE and HOFFMANN 1984) and C9 binding (HÄNSCH et al. 1981). Investigations of the mechanisms of species restriction have been very important in defining the factors responsible for the limitation of membrane damage by homologous complement.

1.3 Isolation of Membrane Inhibitors of Complement

The first description of membrane regulators of complement came from the work of HOFFMANN (1969a), who assayed extracts of solubilized human erythrocyte membranes for their ability to inhibit guinea pig complement lysis of sheep erythrocytes. HOFFMANN distinguished an extract made in the presence of 500 mM sodium phosphate as "inhibitor-high" (I-H) and a second extract made in the presence of 5 mM sodium phosphate as "inhibitor-low" (I-L) (HOFFMANN 1969b). About 20% of I-H activity could bind to sheep E and inhibit subsequent lysis, and, in retrospect, this fraction could have contained significant amounts of glycan phosphatidylinositol (GPI)-anchored proteins, including the membrane regulators of the terminal complement sequence, CD59 and the C8 binding protein, and DAF. In a later paper, (in which) I-H is better characterized, I-H had no effect on the decay of an EAC4b2a intermediate, indicating there was no significant DAF (HOFFMANN et al. 1974). The I-L, by contrast, specifically acted on intermediates bearing C2 but did not incorporate into cell membranes. In retrospect, most of the I-L activity was probably due to complement receptor type I (CR1), which acts at the same site as DAF but does not reincorporate into membranes. HOFFMANN and ETLINGER (1973) also investigated the ability of 0.5 M NaCl extracts of erythrocyte stroma prepared from different species to inhibit an EAC14b2a intermediate made with guinea pig complement. Extracts from cow, nurse shark, horse, chicken, and dog had no effect, while guinea pig extracts were most active and rabbit extracts had some activity. This demonstration of species specificity of complement regulation (species restriction) was almost certainly due to what is now known as DAF.

The first usage of the term "decay accelerating factor" was by OPFERKUCH et al. (1971). They were studying serum factors that accelerated the decay of the EAC4b2a intermediate, which, in retrospect, were what are now termed C4 binding protein and factor H. HOFFMANN then applied the term DAF to his I-L factor, which, as mentioned, was probably CR1.

Over a decade later, with the use of detergent containing buffers and sequential chromatography, including hydrophobic chromatography, it was possible to purify a membrane DAF from guinea pig erythrocytes using as an assay the inhibition of EAC1, $4b^{gp}$, $2a^{gp}$ intermediates (NICHOLSON-WELLER et al. 1981), and to purify a membrane DAF from human erythrocytes, using as an assay the inhibition of EAC1,4b^{hu}, 2a^{gp} intermediates (NICHOLSON-WELLER et al. 1982). Guinea pig DAF was characterized as a single polypeptide chain of 60 kDa and human DAF as a single polypeptide chain of 70 kDA (NICHOLSON-WELLER et al. 1981, 1982).

2 Function of DAF

2.1 DAF Inhibits C3 Activation

C3 activation is a pivotal step in the complement pathway for the following reasons: (1) it occurs at the juncture of the classical and alternative pathways; (2) the generation of cleaved C3 is crucial for the recruitment of the amplification pathway; and (3) the most potent complement mediators, including anaphylatoxins, chemotaxins, and the membrane attack complex, are derived from C3 activation or steps more distal in the pathway. The specificity of DAF's action at the C3 activation step could be partially inferred by its effect on C4b2a, the complement intermediate used to assay DAF during its purification. When DAF is added immediately after the formation of EAC4b2a, the efficiency of C3, C5, C6, C7, C8, and C9 is not affected. However, with time there is a temperature-dependent decay of the C4b2a enzyme and the presence of DAF accelerates this decay (NICHOLSON-WELLER et al. 1981, 1982). Since C4b is covalently bound to the erythrocyte membrane, the decay was assumed to represent the dissociation of C2a. Later, it was appreciated that membrane-isolated DAF, when mixed with target erythrocytes, can reincorporate into the target erythrocyte membrane and function to inhibit complement (MEDOF et al. 1984). In retrospect, in standard assays of purified DAF, most of the DAF activity is due to reincorporated DAF.

DAF also inhibits C3 activation when it is present at the time of C4b2a enzyme formation. Since DAF is very efficient at dissociating C4b2a, it is impossible from simple complement inhibition studies to assess an effect on the formation of C4b2a. However, evidence that DAF does not affect the binding of C2 to C4b but only accelerates the dissociation of the C2a fragment came from studies analyzing the state of radiolabeled C2 found on the surface and in the supernatant of EAC1, 4b, 2a and E^{DAF}AC1, 4b, 2a. At any given time, the amounts of uncleaved C2 on the erythrocyte intermediates are similar, which indicates that the binding of C2 is not effected by the presence of DAF. However, there is always less C2a on the erythrocyte intermediate bearing incorporated DAF and more of the decayed form of C2a in the supernatant (Fig. 1). These results can best be explained by DAF affecting the dissociation of C2a, thereby allowing a more rapid turnover of native C2. Similar findings are obtained with the alternative pathway C3 convertases of EAC4b, 3b, Bb and E^{DAF}C4b, 3b, Bb: DAF does not affect the amount of uncleaved B that is bound, but causes the

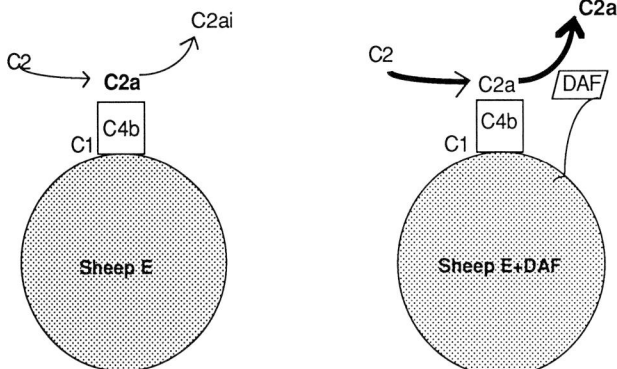

Fig. 1. The consequences of DAF action. *Left*, A sheep erythrocyte with a C14b attached to the membrane. C2 binds to C4b and is activated to form the C3 cleaving enzyme, C4b2a, in which C2a is the catalytic subunit. C4b2a decays by the "spontaneous" dissociation of the C2a subunit, which decays to C2ai, an enzymatically inactive form that cannot rebind to C4b. *Right*, A sheep erythrocyte which has incorporated human DAF in its membrane. The activity of DAF is to accelerate the dissociation of the C2a subunit, resulting in increased turnover of C2 and an accumulation of C2ai, depicted by the *heavy arrows*. See Sect. 2.1. (Based on data from FUJITA et al. 1987)

accelerated release of Bb into the supernatant (FUJITA et al. 1987). Nephritic factors, which are pathologic antibodies with specificity for the C4b2a enzyme or the C3bBb enzyme, are able to abnormally stabilize these C3 activating bimolecular enzymes. The presence of nephritic factor bound to the C4b2a enzyme can prevent decay by DAF. A nephritic factor with specificity for the C3bBb enzyme rendered the enzyme partially resistant to DAF (ITO et al. 1989). Nephritic factors are known to have varying affinities, like all antibodies, and the relative difference in resistance to DAF of C4b2a and C3bBb may not be generalizable if other nephritic factors are tested.

DAF has no cofactor activity for the cleavage of C3b or C4b by factor I (PANGBURN et al. 1983; ROBERTS et al. 1985). This is in contrast to the other membrane (CR1, membrane cofactor protein) and fluid phase (H, C4 binding protein) regulators of the C3 convertases which do act as cofactors for I.

For erythrocytes, DAF is identified as the species restricting factor acting at the C3/C5 activation step (SHIN et al. 1986). DAF has no activity at the C8/C9 step where two other membrane factors, namely, CD59 (DAVIES et al. 1989; HOLGUIN et al. 1989; OKADA et al. 1989; SUGITA et al. 1988) and the C8 binding protein (homologous restriction factor), act (SCHONERMARK et al. 1986; ZALMAN et al. 1986). For monocytes, lymphocytes, and certain cell lines, species restriction operates only at the step of C3 activation and DAF is the responsible membrane factor. Granulocytes are apparently an exception to nucleated cells and do require both restriction at the C8/9 step and C3 restriction by DAF (YAMAMOTO et al. 1990).

2.2 Binding Ligand for DAF

Several lines of evidence indicate that DAF has some affinity for C4b and C3b, but when the catalytic subunits C2a and Bb, respectively, bind, DAF's affinity is enhanced. If DAF had a higher affinity for membrane bound C4b or C3b, its dissociation and hence mobility would be retarded and it would be less efficient at inhibiting the C4b2a and C3bBb enzymes (Fig. 2). The enhanced affinity of DAF for the intact enzymes would explain how the relatively modest number of DAF molecules (3300/erythrocyte; 85 000/granulocyte) can keep recycling over the surface of the cell membrane (KINOSHITA et al. 1985).

Solubilized DAF does not bind to chemically insolubilized C4/C4b and C3/C3b over a wide range of pH and ionic strengths, suggesting that the interaction is weak (NICHOLSON-WELLER, unpublished results). Cross-linking studies with the homobifunctional reagent dithiobis-succinimidyl propionate (DSP) have been used to analyze the interactions of DAF in the membrane. Radiolabeled DAF was reincorporated into sheep erythrocytes bearing either C4b or C3b, and, after cross-linking, the anti-DAF immunoprecipitates were analyzed by SDS PAGE. Alternatively, the C3 cleaving enzymes were assembled on human erythrocytes and, after cross-linking, the complexes were blotted, and the blots probed with anti-DAF or anti-C4. In both cases, DAF was cross-linked to C4b or to C3b (KINOSHITA et al. 1986). Although results from chemical cross-linking studies need cautious interpretation, there are theoretically more reasons why real binding might not be detected than there are reasons why apparent specific binding might be an artifact. Also in favor of DAF binding to C3b or C4b is the fact that DAF contains the short consensus repeats of amino acids, which

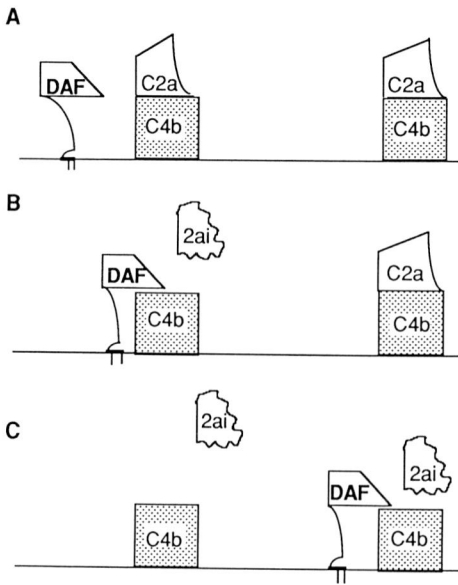

Fig. 2 A–C. The GPI anchor of DAF allows it to move in the plane of the plasma membrane. **A** DAF moves randomly in the membrane until it encounters the bimolecular enzyme C4b2a. **B** DAF dissociates the catalytic subunit C2a and thereby inactivates the C3 cleaving enzyme. **C** Since DAF apparently has a higher affinity for the bimolecular enzyme than for either C4b or C3b, it can recycle to another C4b2a convertase, which it inactivates by dissociation. Although only the classical C3 convertase, C4b2a, is shown, DAF is also active against the alternative pathway C3 convertase, C3bBb, which it inactivates by dissociating the catalytic subunit Bb from the membrane bound subunit C3b. See Sects. 2.1 and 5.3

are characteristic of the other members of the family of regulators of complement activation which bind to C3b and C4b, namely, factor H, C4 binding protein, and CR1.

Experimental evidence that DAF binding is influenced by the presence of the catalytic subunits comes from direct measurements of interactions in the fluid phase. The affinity (apparent K_a or association constant) of DAF for C4b is 0.450 nM^{-1}, while for C4b2a enzyme it is 530 nM^{-1}. There is a similar enhanced affinity of DAF for C3bBb (910 nM^{-1}) compared with C3b (45 nM^{-1}) or Bb (67 nM^{-1}). Also, some affinity of DAF for Bb alone has been measured in the fluid phase (67 nM^{-1}) (PANGBURN 1986). There are at least two possible ways the catalytic subunits may influence DAF binding: (1) the bound catalytic subunits may affect the confirmation of the membrane bound C4b and C3b or (2) there may be an additional binding site on DAF for the bound catalytic subunit. A site on C2a or Bb would convey specificity for the assembled enzymes since these catalytic fragments have unique determinants which are lost when they dissociate from their respective membrane bound subunits.

2.3 Distinct Role of DAF in Regulating C3 Activation at the Membrane

For erythrocytes, DAF and CR1 (C3b receptor or CD35) are the only identified membrane regulators of C3 activation. Interestingly, these two molecules have divided responsibilities for regulating C3 activation: DAF acts on C3 convertases formed on autologous surfaces, while CR1 acts on C3 convertases on the adjacent surfaces of cells or fluid phase immune complexes (MEDOF et al. 1984; ROBERTS et al. 1985). DAF has only four short consensus repeats in its structure, whereas CR1 has 30, which theoretically would enable it to reach neighboring surfaces. The membrane cofactor protein (MCP) is found on nonerythroid cells, and its function is limited to acting as a cofactor for the cleavage of C3b by factor I (reviewed in LUBLIN and ATKINSON 1990).

3 DAF Gene

3.1 Localization

The *DAF* gene was initially cloned from cDNA libraries from HeLa cells (CARAS et al. 1987a) and HL-60 cells (MEDOF et al. 1987a) using oligonucleotide probes based on the NH_2 terminus DAF sequence. Linkage studies of RFLPs of genomic DNA (REY-CAMPOS et al. 1987), somatic cell hybrids, and in situ hybridization (LUBLIN et al. 1987) localized the gene to the long arm of chromosome 1, band q32; it is part of a larger family of genes known as regulators of complement activation (RCA). The RCA region includes the genes for factor H, C4 binding protein, MCP, CR1, CR2, and DAF. Linkage studies suggested that the gene for factor H is separated from the rest of the RCA gene cluster, which has been shown by pulsed field gradient gel electrophoresis to exist in the following order: CR1, CR2, DAF, and C4 binding protein (CARROLL et al. 1988). There is only one copy of the DAF gene (STAFFORD et al. 1988).

3.2 Organization

Two forms of DAF protein were initially predicted from the sequences cloned from the cDNA library made from HeLa cells: a protein of 381 amino acids with a hydrophobic COOH terminus translated from a spliced mRNA species and a protein of 440 amino acids with a hydrophilic COOH terminus translated from an unspliced mRNA species (Fig. 3). The spliced mRNA transcript is ninefold more prevalent than the unspliced mRNA transcript. Spliced and unspliced mRNA transcripts are found on HeLa cell polysomes indicating that they are both translated. In transfected Chinese hamster ovary cells the hydrophobic protein product is expressed on the surface, while the hydrophilic product remains intracellular (CARAS et al. 1987). If was initially believed that the nonspliced hydrophilic form of DAF, which lacks a putative transmembrane domain, might be a secreted form of DAF. To date, a limited search with antibodies directed against the predicted peptide sequence for the hydrophilic form has not detected antigen (I. CARAS, personal communication).

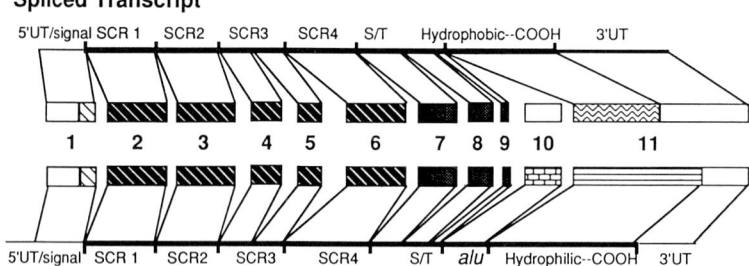

Fig. 3. There is one copy of the *DAF* gene but an alternate splice site allows two different transcripts (*top* and *bottom*). The 11 exons are represented by *boxes*. *Top*, the spliced transcript encodes a 381 amino acid peptide with a hydrophobic COOH terminus for subsequent proteolytic cleavage of the terminal 27 amino acids and addition of the GPI anchor. The splicing allows a frameshift in reading exon 11, which makes the encoded protein domain hydrophobic. Only spliced transcripts have been recognized ex vivo. *Bottom*, the unspliced transcript encodes a 440 amino acid peptide which contains an Alu sequence (exon 10). Exon 11 is not frameshifted and predicts a hydrophobic COOH terminus. This unspliced transcript has been found in cell lines but has not, as yet, been identified in normal or abnormal tissue ex vivo. See Sect. 3.2. (Based on data from CARAS et al. 1987a; MEDOF et al. 1987a; POST et al. 1990)

It is now known that the *DAF* gene (Fig. 3) is organized into 11 exons which encode the following sequences: exon 1, a 5'-UT/signal peptide; exon 2, short consensus repeat I (SCR I), exon 3, SCR II; exons 4–5, SCR III; exon 6, SCR IV; exons 7–9, a serine/threonine rich region; exon 10, an Alu sequence which is spliced out of the predominant GPI-anchored form of DAF; and exon 11 which encodes the COOH terminus of both the spliced and unspliced transcripts of the gene (POST et al. 1990). A frameshift in the transcription of exon 11 predicts a hydrophobic COOH sequence for the spliced transcript, while, for the unspliced transcript, exon 11 is not frameshifted and the transcript encodes a hydrophilic COOH terminus (CARAS et al. 1987a; POST et al. 1990).

Upstream of the NH_2 terminus a 34 amino acid leader sequence is predicted; downstream of the COOH terminus there are four different polyadenylation sites, predicting different sized mRNA transcripts. Two or three major DAF mRNA species have been reported from northern blot analyses of HeLa and HL-60 RNA (CARAS et al. 1987a; MEDOF et al. 1987a).

4 Structure of DAF

4.1 Predicted Peptide Structure

Since the hydrophilic (unspliced) gene product has not yet been identified in normal tissues, the remaining discussion will focus on the hydrophobic species of DAF. Amino acid 1 starts the first of four contiguous 60 + amino acid SCRs (Fig. 4). These SCRs are homologous to the SCRs found in the other members of the RCA proteins as well as complement proteins factors B and H, C2, C1r, and the noncomplement proteins factor XIII, β_2-glycoprotein I, and the interleukin-2 receptor (reviewed in REID et al. 1986). Each SCR is defined by four cysteines, which in the case of β_2-glycoprotein I (LOZIER et al. 1984) and C4 binding protein (JANATOVA et al. 1989) are folded in an S shape so that cys1–cys3 and cys2–cys4 bonds form. It seems likely that the SCRs in DAF will be bonded in a similar manner. Predictions of DAF structure and analysis of the SCRs of factor H suggest that the SCR region of DAF is organized into antiparallel β-sheets (PERKINS et al. 1988).

After the four consensus repeats in DAF, which end at amino acid 250, there is a serine/threonine rich region that extends from residue 253 to residue 322. The portion of the gene which predicts the COOH terminus of translated DAF is similar to gene sequences encoding the COOH termini of the membrane forms of the variable surface glycoprotein of trypanosomes and Thy-1 antigen.

4.2 Glycan Phosphatidylinositol Anchor

The first evidence that DAF was linked to the membrane by a GPI anchor was the demonstration that DAF could be cleaved from the membrane by a

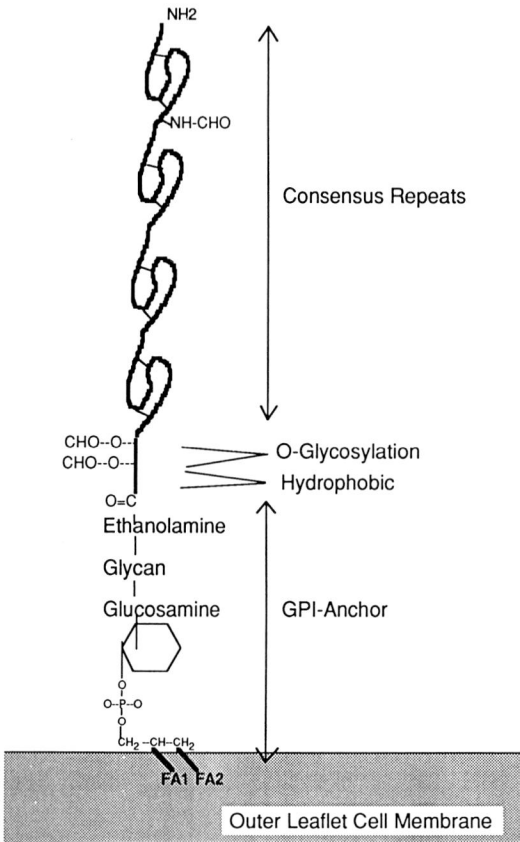

Fig. 4. Structure of DAF. Certain aspects of the DAF molecule were appreciated from the initial efforts to purify and analyze the molecule, namely, is hydrophobicity, single chain structure, and intrachain disulfide bonds. A fuller explanation of these attributes has come from: (1) the predicted peptide sequence from the cloned gene; (2) biosynthetic studies which have given estimates of the type and extent of glycosylation; and (3) the recognition that DAF, like the variable surface glycoprotein of trypanosomes, was anchored to the membrane by a glycan phosphatidylinositol moiety. *FA* = fatty acid; see Sect. 4.0

phosphatidylinositol-specific phospholipase C (PIPLC) (DAVITZ et al. 1986; MEDOF et al. 1986). DAF, when solubilized from butanol extracted erythrocytes and subsequently purified in the presence of detergent, has the capacity to reinsert in membranes. Once cleaved by PIPLC, DAF loses the capacity to reinsert into membranes.

Much of the knowledge of the GPI anchor comes from the study of the variable glycoprotein (VSG) of trypanosomes (reviewed in DOERING et al. 1990), in which GPI-anchored VSG synthesis represents about 10% of the protein synthesized by these organisms. The hydrophobic COOH terminus of the protein is removed and a preformed GPI, glycolipid A, is added by an amide bond forming between the terminal amino acid and the terminal ethanolamine of glycolipid A. These modifications occur in the endoplasmic reticulum. The synthesis of glycolipid A can be carried out in cell-free systems which has allowed identification of the core structure. Glycolipid A is remodeled by changing the fatty acids to myristate just prior to attachment of the anchor in the endoplasmic reticulum. Once the anchor is attached to the protein, the glycan core is modified by the addition of sugar residues before the protein is inserted in the plasma membrane.

The cores of the GPI anchors of DAF and acetylcholinesterase differ in three ways from that of VSG. First, for DAF and acetylcholinesterase, the ratio of saturated to unsaturated fatty acids is about 1, with palmitic and stearic acid comprising the majority of saturated fatty acids and C18:1, C20:4, C22:4, and C22:5 comprising the majority of unsaturated fatty acids. This is in marked contrast to VSG which only contains myristate. Second, one of the diglyceride fatty acids of DAF is attached by an ether linkage, while the second fatty acid is attached by an ester linkage (alkylacylglycerol); in VSG both fatty acids are linked to glycerol by ester bonds (diacylglycerol). Third, the inositol of erythrocyte DAF is substituted with an ester-linked fatty acid. Interestingly, although this modification is seen in erythrocyte acetylcholinesterase, it is not seen in DAF from nonerthyroid sources (WALTER et al. 1990). Thus, there is the potential for cell-specific as well as species-specific modifications of the GPI core.

The sequences directing the removal of the COOH terminus amino acids and the condensation of preformed glycolipid A are included in the COOH region of the protein. This has been shown by fusing the 3' translated end of the *DAF* gene encoding 37 amino acids to the 3' terminus of the translated end of a secreted protein. The resultant fusion protein was expressed in the membrane with a GPI anchor (CARAS et al. 1987b). If the terminal 17 amino acids are removed from DAF, the GPI anchor will not attach to DAF, but the transfer of these amino acids to a secreted protein will not direct the addition of GPI to the fusion protein. Thus, the terminal 17 amino acids of DAF are necessary but not sufficient as a signal for attaching the GPI (CARAS et al. 1989). Interestingly, the sequence of the terminal 17 amino acids directing the GPI posttranslational modification is apparently not specific, since the substitution of a random hydrophobic sequence also provides a signal for the both the addition of a GPI anchor and transport and expression on the plasma membrane (CARAS and WEDDELL 1989). More recently, it has been possible to identify and sequence the tryptic peptide of [^{3}H]ethanolamine biosynthetically labeled DAF. The ethanolamine is attached to serine 319, implying that the 28 more distal amino acids are removed. Modifications more NH_2-terminal to serine 319 do not affect the attachment of the GPI anchor (MORAN et al. 1991).

4.3 Glycosylation of DAF

The predicted protein sequence of the *DAF* gene, after the removal of 28 amino acids from the COOH terminus and the addition of the GPI anchor, would have a molecular weight of about 46 kDa, which is what is seen intracellularly. From the analysis of immunoprecipitates of solubilized surface-iodinated cells that had been treated with various endo- and exoglycosidases, it has been possible to identify the relative size and extent of glycosylation (LUBLIN et al. 1986). Endo F, but not endo H, decreases the mobility of DAF by about 3 kDa, indicating that the one N-glycosylation site at amino acid residue 61 has a complex type carbohydrate of about 3 kDa. The molecular weight of mature DAF is decreased by 18 kDa after neuraminidase treatment and a further 8 kDa after endo-α-N-acetylgalactosaminidase treatment. Sequential digestions indicated that most of the sialic acid is on the O-linked side chains, and the total O-linked carbohydrate accounts for 25 kDa. Biosynthetic studies were performed in differentiated HL-60 cells, and the earliest [^{35}S] methionine precursor to be noted had a molecular mass of 43 kDa. This is rapidly converted into a species with an apparent molecular mass, of 46 kDa, presumably by modifications in the peptide chain and the attachment of the GPI anchor. The O-linked side chains, although not necessary for DAF function, have been shown to be critical for the resistance of the mature membrane DAF against proteolysis (REDDY et al. 1989).

4.4 Alternate Forms of Mature DAF

Additional lower molecular mass forms of DAF have been noted during purification: a 63 kDa species, designated DAF-A, and a 55 kDa species, designated DAF-B (SEYA et al. 1987). DAF-A and DAF-B are thought to represent degraded forms of membrane DAF because they are more prevalent when DAF is purified from the membranes of aged erythrocytes. Both the 55 kDa and the 63 kDa forms are active in regulating fluid phase C3 activating convertases with comparable specific activities to native DAF. They were also tested for their ability to reincorporate into erythrocyte membranes and to regulate the cell bound C3bBb assembled on the cells by acid-activated serum. Compared with native DAF, both lower molecular mass forms were relatively inactive in this assay. The 63 kDa form showed slightly more activity, which is consistent with its retention on a hydrophobic column and its ability to be released by PIPLC digestion of intact erythrocytes. The 55 kDa DAF is not hydrophobic and cannot be released from intact cells by PIPLC, suggesting that its GPI anchor is missing or defective. Papain treatment of intact surface-iodinated erythrocytes releases a 55 kDa DAF antigen, and this release could be duplicated by a 2 h 37°C coincubation of surface-labeled erythrocytes with leukocytes from the same donor. The structures of DAF which are found in the fluid phase, i.e., serum and urine, have not been characterized in detail, but these DAF molecules are only slightly smaller than native DAF and cannot reincorporate into membranes, consistent with their loss of the GPI anchor.

Higher molecular mass forms of DAF, 100 kDa and 140 kDa, have also been described (KINOSHITA et al. 1987; LASS et al. 1990) in blots of nonreduced membrane extracts and in SDS PAGE of reduced radioiodinated DAF. The 140 kDa form has complement inhibitory activity and can reincorporate into membranes. It seems possible that this form is a dimer of the native DAF, which in the gel system used has an apparent electrophoretic mobility of 70 kDa.

5 Expression and Distribution of DAF

5.1 Expression on Cells

DAF has been found on a wide variety of circulating cells and tissues. Among circulating blood cells, DAF is present on erythrocytes, granulocytes, and monocytes with a unimodal distribution by indirect immunofluorescence and FACS analysis (KINOSHITA et al. 1985; NICHOLSON-WELLER et al. 1985a). Lymphocytes, by contrast, have a skewed distribution of DAF expression, with a variable proportion (1%–30%) being apparently DAF deficient. Analysis of DAF expression on lymphocytes from normal donors indicates that about half the DAF deficient population is $CD3^+$ with a mixture of $CD4^+$ and $CD8^+$, while the other half of the DAF deficient lymphocytes are $CD16^+$ NK cell (NICHOLSON-WELLER et al. 1986). All NK cells are uniformly DAF-negative, although they are capable of becoming DAF-positive after culture in vitro (FINBERG and NICHOLSON-WELLER, unpublished observations). Leukocyte DAF has a higher molecular mass than erythrocyte DAF (NICHOLSON-WELLER et al. 1985a), and biosynthetic studies suggest that this is due to increased glycosylation (LUBLIN et al. 1986).

A wide variety of normal tissues express DAF including endothelial cells of the vasculature (ASCH et al. 1986) and endocardium; cells of the uterus, synovia, and urinary and gastrointestinal tracts; and some exocrine glands (MEDOF et al. 1987b; LASS et al. 1990). In tissues with a gradient of differentiated cells, DAF is more prevalent in the most mature cells. In tissues with a polarity to the cells, DAF expression is concentrated on the apical aspect of the cells which comprise the luminal surface of the tissue. This restriction of DAF to the apical cell surface is true for all GPI-linked proteins. Experimentally, fusion of the 37 NH_2 terminus amino acids of DAF, which contain the signal for GPI attachment, with the ectodomain of herpes simplex glycoprotein D, which is normally a basolateral antigen, or with human growth hormone, which is normally secreted, causes both these GPI fusion proteins to be expressed exclusively on the apical surface of the MDCK cell (LISANTI et al. 1989). DAF has also been noted in the extracellular fluids of the eye (LASS et al. 1990) and the extracellular matrix of some tissues including esophagus, uterus, and endocardium. Interestingly, the staining in the extracellular matrix is in a fibrillar pattern (MEDOF et al. 1987b). The structure and function of this DAF is unknown. Little is known about DAF expression during ontogeny. One study shows minimal DAF expression in fetal liver, but exuberant expression within the placenta where the maternal circulation is exposed to fetal tissue (HOLMES et al. 1990).

5.2 Regulation of Membrane Expression

Control of DAF expression at the transcriptional and translational levels has not been reported to date. This information may be hard to collect because, at least for some cells, namely, granulocytes, there is a large intracellular pool of DAF that can be up-regulated by f-met-leu-phe or zymosan-activated serum, both chemoattractants (BERGER and MEDOF 1987). The time course of stimulation is rapid, with half-maximal response in less than 4 min and a plateau by 20 min. During this time, DAF expression on granulocytes doubles from 10 000 to over 20 000 sites/cell. Puromycin and cycloheximide have no effect on the up-regulation of DAF, confirming that protein synthesis is not involved. The time courses of up-regulation of CR1 and CR3 parallel that of DAF. Since the up-regulation of both DAF and CR1 is not inhibited by chelating Ca^{2+} in the medium, whereas the up-regulation on CR3 does require extracellular Ca^{2+}, the DAF and CR1 may be in a compartment distinct from that of CR3. The up-regulation of DAF and CR1 is inhibited by trifluoroperazine, which indicates that intracellular Ca^{2+} is involved in the translocation of these molecules. Circulating T lymphocytes or cloned T cells will increase their DAF expression when stimulated with the mitogen phytohemagglutinin (DAVIS et al. 1988).

The location of the intracellular pools of DAF has been an area under investigation. Although not yet proven, it seems likely that DAF may be stored in a newly described secretory vesicle which contains GPI-anchored alkaline phosphatase (BORREGAARD et al. 1987) and tetranectin (BORREGAARD et al. 1990), a secreted protein. DAF can be down-regulated from the surface of granulocytes by endocytosis. Granulocytes incubated in buffer at 37°C lose 44% of their DAF antigen over a 90 min period, and most of it is internalized. Cross-linking with anti-DAF accelerates endocytosis but is not essential. Internalized DAF is first seen in endocytic vesicles and then later in multivesicular bodies (TAUSK et al. 1989). The potential pathways for recycling DAF are being actively studied.

DAF expression can also be down-regulated by shedding of DAF from membranes. Exposure of erythrocytes to calcium ionophore can lead to the loss of vesicles which are enriched for GPI-linked proteins (BUTIKOFER et al. 1989). Whether this might occur under more physiologic conditions or with nucleated cells is not known. When DAF is cross-linked on the surface of granulocytes using monoclonal anti-DAF and followed by radiolabeling with anti-mouse IgG and incubation, there is a temperature-dependent loss of DAF into the medium: 26% of membrane DAF is lost at 37°C while only 6% is lost at 4°C. The molecular mass of the shed DAF is 56 kDa, while that of DAF that is released by PIPLC cleavage of the GPI anchor is 65 kDa, which suggests that the shed DAF is proteolytically cleaved (TAUSK et al. 1989).

There is no evidence to date that DAF is normally shed from the membrane by cleavage of the GPI anchor. Mammalian PIPLCs are well described, but they are cytosolic or associated with membranes inside the cell (reviewed in RHEE et al. 1989). A mammalian serum phosphatidylinositol-specific phospholipase D has been described (DAVITZ et al. 1987); its role in the metabolism of GPI-anchored proteins is unknown.

Bacteria can also produce a PIPLC, and if the bacterium has invasive characteristics, such as *Staphylococcus aureus*, the PIPLC secreted during the course of an infection could potentially cleave GPI-linked proteins from the surface of host cells. The loss of the complement regulatory proteins DAF, C8 binding protein, and CD59 would augment inflammation by allowing host complement to damage host cells. In a small study of S. aureus infections, those patients with disseminated intravascular coagulopathy and adult respiratory distress syndrome, conditions associated with unregulated complement activation, were more likely to be infected with strains producing PIPLC than were patients without these clinical syndromes (MARQUES et al. 1989).

5.3 Mobility in the Membrane

A consequence of DAF having a GPI anchor is that it is very mobile in the membrane. Photobleaching data indicate that DAF, when tagged with fluoresceinated intact antibody, moves an estimated 1.6×10^{-9} cm^2/s, while glycophorin A, which is a transmembrane protein, moves about 0.18×10^{-9} cm^2/s (THOMAS et al. 1987). The rapid mobility of DAF would compensate for the fact that relatively few molecules must be able to control complement activation over the entire cell surface. Thus, the estimated 3000 DAF molecules per erythrocyte (KINOSHITA et al. 1985) means that each DAF molecule is responsible for about 45×10^{-9} cm^2 of membrane. Chemical cross-linking experiments indicate that DAF is present as a monomer in the plasma membrane (KAMMER et al. 1988).

5.4 Can DAF Transmit Transmembrane Signals?

There has been a great deal of interest as to whether the GPI anchors of proteins can be utilized to transmit intracellular messages (reviewed in ROBINSON 1991). Much of this work has been based on studies of GPI-anchored Thy-1, a lymphocyte activation antigen of the rat. Cross-linking of Thy-1 by specific antibodies leads to an increase in intracellular Ca^{2+} and subsequent secretion of IL-2. It was thought initially that the signaling might utilize the phosphatidylinositol of the anchor, but now there is evidence that cross-linking may perturb a neighboring T cell receptor, and it is the T cell receptor which is responsible for transmitting the intracellular signal (ASHWELL and KLUSNER 1990). When either the T cell receptor or Thy-1 is specifically immunoprecipitated from surface-iodinated lymphocytes, CD45, a tyrosine phosphatase, is coimmunoprecipitated (VOLAREVIC et al. 1990). CD45 may modulate T cell receptor signaling when cross-linked GPI-anchored Thy-1 causes an association or perturbs the association of certain transmembrane proteins.

There are two studies of the effects of anti-DAF cross-linking. First, when DAF is cross-linked on the surface of freshly isolated T lymphocytes using anti-DAF, DAF caps and there is an accumulation of cytoskeletal elements beneath the cap. The capping process is facilitated by colchicine and inhibited by cytochalasins B and D. In the second study, the effect of cross-linked DAF as a signal for proliferation was tested on peripheral blood human T cells. DAF must be cross-linked with two antibodies, first, a polyclonal rabbit anti-DAF and second, goat anti-rabbit Ig; then, if a substimulatory dose of phorbol ester is also used, the T cell proliferate (DAVIS et al. 1988). Although IL-2 could augment the response, phorbol ester was required. It seems possible that the extensive cross-linking necessary in these experiments to get an anti-DAF effect might affect neighboring transmembrane proteins that in turn do the signaling, as has been shown for Thy-1 (vide supra). DAF then may function as a coreceptor for complement-bearing ligands. To data there are no data implicating use of the phosphatidylinositol within the GPI anchor as substrate for the intracellular phosphatidylinositol signaling system.

5.5 DAF in Extracellular Tissues

Soluble DAF antigen has been detected and quantitated by ELISA in the following fluids from normals: plasma (64 ng/ml), tears (344 ng/ml), saliva (112 ng/ml), synovial fluid (168 ng/ml), and cerebral spinal fluid (38 ng/ml) (MEDOF et al. 1987b). The DAF antigen of tears and saliva has a molecular mass of > 100 kDa compared with 70 kDa of erythrocyte membrane DAF. DAF is also found in urine, with a 24 h excretion ranging from 150 to 510 µg for 11 donors. Urine DAF is not able to incorporate into membranes, has a molecular weight 3 kDa less than erythrocyte membrane DAF, and is able to function in the fluid phase as an inhibitor of EAC1, 4b, 2a. These data are consistent with urine DAF lacking a GPI anchor, but there is no direct proof that urine DAF represents a shed form of membrane DAF nor any information as to its cell of origin.

6 Abnormal Expression of DAF

6.1 Inab Phenotype

Antibodies recognizing the Cromer-related antigens were found to react with a 70 kDa protein of erythrocyte membranes which had been western blotted (SPRING et al. 1987). Subsequently, it was shown that the 70 kDa protein was DAF (TELEN et al. 1988). Four individuals lacking the Cromer antigens have been identified and were designated as having an Inab phenotype (DANIELS et al. 1982; WALTHERS et al. 1983). Intact Inab erythrocytes bind minimal amounts of anti-DAF antibody, and western blotted Inab erythrocyte membranes do not react with anti-DAF. Thus, Inab cells are apparently lacking the DAF molecule. Functionally, patients with Inab cells do not have frank hemolysis, although there are no reported data on red cell survival. In vitro Inab erythrocytes demonstrate an abnormal sensitivity to the sucrose lysis test, in which the alternative complement pathway is activated, and the complement lysis sensitivity assay, in which the classical complement pathway is activated by antibody. When a membrane bound C3bBb enzyme was assembled on normal erythrocytes and Inab erythrocytes and the amount of membrane bound C3 antigen compared, significantly more C3 was activated and bound on Inab erythrocytes (TELEN and GREEN 1989). This finding is consistent with Inab cells being functionally deficient in DAF and therefore unable to regulate C3 activation at the membrane. Interestingly, Inab erythrocytes are not lysed in the acidified serum lysis or Ham test (MERRY et al. 1989; TELEN and GREEN 1989). The Ham test is the classic clinical test for diagnosing paroxysmal nocturnal hemoglobinuria (PNH) and it relies on complement activation both by the alternative pathway (GÖTZE and MÜLLER-EBERHARD 1972) and by directly activating C5 to form C5–9 (HÄNSCH et al. 1983). It has not been clear which mode of activation is more important for lysing PNH cells, but the fact that Inab (DAF deficient) cells are not lysed in the Ham test suggests that the Ham test may stress membrane defenses against C5–9 more than defenses against C3 deposition.

6.2 Paroxysmal Nocturnal Hemoglobinuria

PNH is a disease caused by an acquired mutation in a bone marrow progenitor cell which gives rise to abnormal clones of erythrocytes, monocytes, granulocytes, platelets, and sometimes lymphocytes (ASTER and ENRIGHT 1969). PNH cells were found to be deficient in DAF (PANGBURN et al. 1983; NICHOLSON-WELLER et al. 1983, 1985; KINOSHITA et al. 1985) and other membrane proteins requiring a GPI anchor for attachment to the cell membrane (reviewed in HALPERIN and NICHOLSON-WELLER 1989). The mutation in PNH presumably blocks the synthesis of the GPI anchor or blocks the assembly of the cellular cofactors necessary for the attachment of the anchor to the proteins, since the DAF gene and its mRNA transcripts are normal in PNH cells (STAFFORD et al. 1988). In addition to DAF, the two other complement regulatory proteins which protect the autologous membrane surface from complement attack, CD59 (DAVIES et al. 1989; HOLGUIN et al. 1989; OKADA et al. 1989; SUGITA et al. 1988), which restricts the binding of C5b-9, and the C8 binding protein or homologous restriction factor (HÄNSCH et al. 1987, 1988; ZALMAN et al. 1987), which limits the interaction of C9 with C5b-8, are also GPI-linked. The clinical syndrome of PNH is characterized by intermittent hemolysis and frequently, thrombotic episodes and marrow hypoplasia. The hemolysis is due to an abnormal sensitivity of PNH erythrocytes to complement-mediated lysis and can be directly related to their deficiencies of DAF, C8 binding protein, and CD59. By virtue of their complement sensitivity, distinct types of erythrocytes have been identified in PNH patients (ROSSE 1973): type I PNH erythrocytes have a normal sensitivity toward complement; type II cells, when sensitized with antibody, are lysed by dilutions of serum in the range of 1/2–1/4. Type II cells are not lysed by purified activated C5b-9, the membrane attack complex of complement. Type III erythrocytes, when sensitized with antibody, are lysed with serum dilutions in the range of 1/20–1/25. Type III cells can be lysed by activated C5b-9. Phenotypically, type I cells are normal: type II PNH erythrocytes are probably partially deficient in GPI anchored proteins (NICHOLSON-WELLER et al. 1985b), although there is the possibility that they are preferentially deficient in DAF as compared with C8 binding protein and CD59 (EDBERG et al. 1991). Type III cells are functionally deficient in all the complement regulatory GPI-anchored proteins, implying that they have a complete block in GPI synthesis. Individual PNH patients may have mixtures of the various abnormal types of cells and normal cells. Investigations of PNH erythrocytes have been very helpful in defining the roles that DAF, CD59, and the C8 binding protein have in protecting autologous membranes of normal erythrocytes from complement-mediated damage.

6.3 Expression in Other Diseases

Expression of DAF has been associated with resistance to antibody-initiated complement-mediated tumor cell lysis. Two melanoma cell lines, one with no DAF and the other expressing high amounts of DAF, were sensitized with anti-ganglioside antibody and then exposed to human serum. The cell lines expressing DAF were relatively resistant to complement-mediated lysis as compared with the DAF deficient cell line. The DAF-positive line could be rendered susceptible to complement-mediated lysis if the DAF was blocked by anti-DAF IgG or Fab_2. The DAF deficient line could be rendered resistant to complement-mediated lysis if exogenous DAF was incorporated into cell membranes. Using the same sensitizing antibody, DAF expression did not affect antibody-dependent cell-mediated cytotoxicity (ADCC) (CHEUNG et al. 1988). Antibody and complement can inhibit tumor growth in vivo. Just how important DAF expression is in enabling tumor cells to escape this mechanism of immune surveillance remains to be seen.

The peripheral blood mononuclear cells from patients with acquired immunodeficiency syndrome (AIDS), when sensitized with antibody, are relatively more susceptible to complement-mediated lysis than the mononuclear cells from normal individuals. The greater susceptibility of mononuclear cells from AIDS patients was correlated with lower DAF expression; the susceptibility could be overcome by the incorporation of exogenous DAF into target cell membranes. There was no preponderance of DAF deficient cells among $CD4^+$ or $CD8^+$ lymphocytes (LEDERMAN et al. 1989). It is not clear whether the relative DAF deficiency of lymphocytes from AIDS patients may contribute to the lymphopenia characteristic of this disease.

7 DAF in Nonhuman Species

7.1 Mammals

DAF has been purified from guinea pig (NICHOLSON-WELLER et al. 1981), rabbit (SUGITA et al. 1987), and mouse erythrocytes (KAMEYOSHI et al. 1989). In each of these species the molecule is a single polypeptide chain of 60–70 kDa, has hydrophobic properties, is widely distributed on various cells within the species, and can decay the C4b2a enzyme. Rat glomeruli possess a membrane factor which is species restricting at the C3 step and is sensitive to trypsin, pronase, and PIPLC. This is the rat homologue of human DAF, which is found on human glomerular epithelial cells (QUIGG et al. 1989). DAF is a species restricting factor, and it is not predictable how efficient DAF will be as a regulator of complement from a particular heterologous species. Human DAF is about 100-fold less efficient in regulating $C4^{gp}$, 2^{gp} convertases than $C4^{hu}$, 2^{gp} convertases. It is likely, although unproven, that sheep erythrocytes, which are the standard target cells for human and guinea pig complement titrations, possess DAF. It is sheep DAF that is responsible for what has always been considered the "intrinsic decay" of human and guinea pig C4b2a convertases seen in standard assays. Rabbit anti-human DAF is poorly cross-reactive with DAF of other species.

7.2 Parasites

A major goal of parasites is to circumvent being recognized as foreign and thereby evade triggering the normal mediators of inflammation. Evading complement activation is part of that scheme. Two interesting examples of DAF expression in parasites have been described. First, in *Trypanosoma cruzi* there is expression of a DAF homologue in the trypomastigotes, a serum resistant stage of the parasite which is normally exposed to human serum. There is no DAF expression in the serum sensitive epimastigote, a stage of *T. cruzi* completed within the insect vector. Trypanosome DAF can accelerate the decay of both the classical and alternative pathway C3 convertases and it has no I cofactor activity (RIMOLDI et al. 1988). This DAF has been purified and characterized as a 87–90 kDa glycoprotein (JOINER et al. 1988).

The second example of DAF on the surface of a parasite occurs in *Schistosoma mansoni*. Schistosomes recovered from experimental animals do not activate the complement of their host species but will activate complement from other species. This species restricted activation suggested that a DAF-like factor might be involved. Anti-human DAF can immunoprecipitate a 70 kDa protein from the surface of schistosomes obtained from primates. In addition, anti-DAF immunofluorescence can be partially removed by PIPLC (PEARCE et al. 1990). Taken together, these data suggest that schistosomes can capture intact functional DAF from their host and incorporate the DAF into their own membranes.

8 Concluding Remarks

The recognition of species restricting factors, the identification of DAF activity in the membrane, the purification and cloning of DAF, the recognition of DAF as a member of the RCA family, and the definition of the function of DAF all represent the fruits of a productive cross-fertilization between basic science and clinical studies. Future studies of DAF will undoubtedly extend our appreciation of the role of complement in host defenses and inflammation and may also shed light on some fundamental biologic questions, such as the way the RCA gene cluster arose from reduplication, the molecular basis and evolutionary implications of species restriction, and the role of GPI-anchored proteins as coreceptors.

Acknowledgements. P.F. Weller, R. Jack, and J.P. Laclette are thanked for reviewing the manuscript. Supported by a grant from the National Institutes of Health, HL33768.

References

Asch AS, Kinoshita T, Jaffe EA, Nussenzweig V (1986) Decay-accelerating factor is present on cultured human umbilical vein endothelial cells. J Exp Med 163: 221–226

Ashwell JD, Klusner RD (1990) Genetic and mutational analysis of the T-cell antigen receptor. Annu Rev Immunol 8: 139–167

Aster R, Enright S (1969) A platelet and granulocyte membrane defect in paroxysmal nocturnal hemoglobinuria: usefulness for the detection of platelet antibodies. J Clin Invest 48: 1199–1210

Berger M, Medof ME (1987) Increased expression of complement decay-accelerating factor during activation of human neutrophils. J Clin Invest 79: 214–220

Borregaard N, Miller LJ, Springer TA (1987) Chemoattractant-regulated mobilization of a novel intracellular compartment in human neutrophils. Science 237: 1204–1206

Borregaard N, Christensen L, Bjerrum OW, Birgens HS, Clemmensen I (1990) Identification of a highly mobilizable subset of human neutrophil intracellular vesicles that contain tetranectin and latent alkaline phosphatase. J Clin Invest 85: 408–416

Butikofer P, Kuypers FA, Xu CM, Chiu DT, Lubin B (1989) Enrichment of two glycosyl-phosphatidylinositol-anchored proteins, acetylcholinesterase and decay accelerating factor, in vesicles released from human red blood cells. Blood. 74: 1481–1485

Caras IW, Weddell GN (1989) Signal peptide for protein secretion directing glycophospholipid membrane anchor attachment. Science 243: 1196–1198

Caras IW, Davitz MA, Rhee L, Weddell G, Martin DW Jr, Nussenzweig V (1987a) Cloning of decay-accelerating factor suggests novel use of splicing to generate two proteins. Nature 325: 545–548

Caras IW, Weddell GN, Davitz MA, Nussenzweig V, Martin DJ (1987b) Signal for attachment of a phospholipid membrane anchor in decay accelerating factor. Science 238: 1280–1283

Caras IW, Weddell GN, Williams SR (1989) Analysis of the signal for attachment of a glycophosph-olipid membrane anchor. J Cell Biol 108: 1387–1396

Carroll MC, Alicot EM, Katzman PJ, Klickstein LB, Smith JA, Fearon DT (1988) Organization of the genes encoding complement receptors type 1 and 2, decay-accelerating factor, and C4-binding protein in the RCA locus on human chromosome 1. J Exp Med 167: 1271–1280

Cheung NK, Walter EI, Smith MW, Ratnoff WD, Tykocinski ML, Medof ME (1988) Decay-accelerating factor protects human tumor cells from complement-mediated cytotoxicity in vitro. J Clin Invest 81: 1122–1128

Daniels GL, Tohyama H, Uchikawa M (1982) A possible null phenotype in the Cromer blood group complex. Transfusion 22: 362–363

Davies A, Simmons DL, Hale G, Harrison RA, Tighe H, Lachmann PJ, Waldmann H (1989) CD59, an LY-6-like protein expressed in human lymphoid cells, regulates the action of the complement membrane attack complex on homologous cells. J Exp Med 170: 637–654

Davis LS, Patel SS, Atkinson JP, Lipsky PE (1988) Decay-accelerating factor functions as a signal transducing molecule for human T cells. J Immunol 141: 2246–2252

Davitz MA, Low MG, Nussenzweig V (1986) Release of decay-accelerating factor (DAF) from the cell membrane by phosphatidylinositol-specific phospholipase C (PIPLC). Selective modification of a complement regulatory protein. J Exp Med 163: 1150–1161

Davitz MA, Hereld D, Shak S, Krakow J, Englund PT, Nussenzweig V (1987) A glycan-phosphatidylinositol-specific phospholipase D in human serum. Science 238: 81–84

Doering TL, Masterson WJ, Hart GW, Englund PT (1990) Biosynthesis of glycosyl phophatidylinositol membrane anchors. J Biol Chem 265: 611–614

Edberg JC, Salmon JE, Whitlow M, Kimberly RP (1991) Preferential expression of human Fc gammaRIIIPMN (CD16) in paroxysmal nocturnal hemoglobinuria. J Clin Invest 87: 58–67

Fries LF III, Frank MM (1987) Molecular mechanisms of complement action. In: Stamatoyannopoules. Nienhuir AW, Leder P, Majerus PW (eds) The molecular basis of blood diseases. Saunders, Philadelphia, chap 13

Fujita T, Inoue T, Ogawa K, Iida K, Tamura N (1987) The mechanism of action of decay-accelerating factor (DAF). DAF inhibits the assembly of C3 convertases by dissociating C2a and Bb. J Exp Med 166: 1221–1228

Götze O. Müller-Eberhard HJ (1972) Paroxysmal nocturnal hemoglobinuria. Hemolysis initiated by the C3 activator system. N Engl J Med 286: 180–184

Halperin JA, Nicholson-Weller A (1989) Paroxysmal nocturnal hemoglobinuria. A complement-mediated disease. Compl Inflamm 6: 65–72

Hänsch G, Hammer CH, JiJi R, Rother U, Shin ML (1983) Lysis of paroxysmal nocturnal hemoglobinuria erythrocytes by acid-activated serum. Immunobiology 164: 118–126

Hänsch GM, Hammer CH, Vanguri P, Shin ML (1981) Homologous species restriction in lysis of erythrocytes by terminal complement proteins. Proc Natl Acad Sci USA 78: 5118–5121

Hänsch GM, Schönermark S, Roelcke D (1987) Paroxysmal nocturnal hemoglobinuria type III. Lack of an erythrocyte membrane protein restricting the lysis by C5b-9. J Clin Invest 80: 7–12

Hänsch GM, Weller PF, Nicholson-Weller A (1988) Release of C8 binding protein (C8bp) from the cell membrane by phosphatidylinositol-specific phospholipase C. Blood 72: 1089–1092

Hoffmann E (1969a) Inhibition of complement by a substance isolated from human erythrocytes I. Extraction from human erythrocyte stromata. Immunochemistry 6: 391–403

Hoffmann E (1969b) Inhibition of complement by a substance isolated from human erythrocytes II. Studies on the site and mechanism of action. Immunochemistry 6: 405–419

Hoffmann E, Etlinger HM (1973) Extraction of complement inhibitory factors from the erythrocytes of non-human species. J Immuno 111: 946–951

Hoffmann E, Cheng W, Tomeu E, Renk C (1974) Resistance of sheep erythrocytes to immune lysis by treatment of the cells with a human erythrocyte extract: studies on the site of inhibition. J Immunol 113: 1501–1509

Holguin MH, Fredrick LR, Bernshaw NJ, Wilcox LA, Parker CJ (1989) Isolation and characterization of a membrane protein from normal human erythrocytes that inhibits reactive lysis of the erythrocytes of paroxysmal nocturnal hemoglobinuria. J Clin Invest 84: 7–17

Holmes CH, Simpson KL, Wainwright SD, Tate CG, Houlihan JM, Sawyer IH, Rogers IP, Spring FA, Anstee DJ, Tanner MJ (1990) Preferential expression of the complement regulatory protein decay accelerating factor at the fetomaternal interface during human pregnancy. J Immunol 144: 3099–3105

Houle JJ, Hoffmann EM (1984) Evidence for restriction of the ability of complement to lyse homologous erythrocytes. J Immunol 133: 1444–1452

Ito S, Tamura N, Fujita T (1989) Effect of decay-accelerating factor on the assembly of the classical and alternative pathway C3 convertases in the presence of C4 or C3 nephritic factor. Immunology 68: 449–452

Janatova J, Reid KBM, Willis AC (1989) Involvement of disulfide bonds in the structure of complement regulatory proteins: C4-binding protein. Biochemistry 28: 4754–4761

Joiner KA, Dias daSilva W, Rimoldi MT, Hammer CH, Sher A, Kipnis TL (1988) Biochemical characterization of a factor produced by trypomastigotes of *Trypanosoma cruzi* that accelerates the decay of complement C3 convertases. J Biol Chem 263: 11327–11335

Kameyoshi Y, Matsushita M, Okada H (1989) Murine membrane inhibitor of complement which accelerates decay of human C3 convertase. Immunology 68: 439–444

Kammer GM, Walter El, Medof ME (1988) Association of cytoskeletal re-organization with capping of the complement decay-accelerating factor on T lymphocytes. J Immunol 141: 2924–2928

Kinoshita T, Medof ME, Silber R, Nussenzweig V (1985) Distribution of decay-accelerating factor in the peripheral blood of normal individuals and patients with paroxysmal nocturnal hemoglobinuria. J Exp Med 162: 75–92

Kinoshita T, Medof ME, Nussenzweig V (1986) Endogenous association of decay-accelerating factor (DAF) with C4b and C3b on cell membranes. J Immunol 136: 3390–3395

Kinoshita T, Rosenfeld SI, Nussenzweig V (1987) A high MW form of decay-accelerating factor (DAF-2) exhibits size abnormalities in paroxysmal nocturnal hemoglobinuria erythrocytes. J Immunol 138: 2994–2998

Lass JH, Walter EI, Burris TE, Grossniklaus HE, Roat MI, Skelnik DL, Needham L, Singer M, Medof ME (1990) Expression of two molecular forms of the complement decay-accelerating factor in the eye and lacrimal gland. Invest Ophthalmol Vis Sci 31: 1136–1148

Lederman MM, Purvis SF, Walter EI, Carey JT, Medof ME (1989) Heightened complement sensitivity of acquired immunodeficiency syndrome lymphocytes related to diminished expression of decay-accelerating factor. Proc Natl Acad Sci USA 86: 4205–4209

Lisanti MP, Caras IW, Davitz MA, Rodriguez BE (1989) A glycophospholipid membrane anchor acts as an apical targeting signal in polarized epithelial cells. J Cell Biol 109: 2145–2156

Lozier J, Takahashi N, Putnam FW (1984) Complete amino acid sequence of human plasma beta 2-glycoprotein I. Proc Natl Acad Sci USA 81: 3640–3644

Lublin DM, Atkinson JP (1990) Decay-accelerating factor and membrane cofactor protein. In: Lambris JD (ed) The third component of complement. Chemistry and biology. Springer, Berlin Heidelberg New York, pp 123–145 (Current topics in microbiology and immunology, vol 153)

Lublin DM, Krsek SJ, Pangburn MK, Atkinson JP (1986) Biosynthesis and glycosylation of the human complement regulatory protein decay-accelerating factor. J Immunol 137: 1629–1635

Lublin DM, Lemons RS, Le BM, Holers VM, Tykocinski ML, Medof ME, Atkinson JP (1987) The gene encoding decay-accelerating factor (DAF) is located in the complement-regulatory locus on the long arm of chromosome 1. J Exp Med 165: 1731–1736

Marques MB, Weller PF, Parsonnet J, Nicholson-Weller A (1989) Phosphatidylinositol-specific phospholipase C (PIPLC), a possible virulence factor of Staphylococcus aureus. J Clin Microbiol 27: 2451–2454

Medof ME, Kinoshita T, Nussenzweig V (1984) Inhibition of complement activation on the surface of cells after incorporation of decay-accelerating factor (DAF) into their membranes. J Exp Med 160: 1558–1578

Medof ME, Walter EI, Roberts WL, Haas R, Rosenberry TL (1986) Decay accelerating factor of complement is anchored to cells by a C-terminal glycolipid. Biochemistry 25: 6740–6747

Medof ME, Lublin DM, Holers VM, Ayers DJ, Getty RR, Leykam JF, Atkinson JP, Tykocinski ML (1987a) Cloning and characterization of cDNAs encoding the complete sequence of decay-accelerating factor of human complement. Proc Natl Acad Sci USA 84: 2007–2011

Medof ME, Walter EI, Rutgers JL, Knowles DM, Nussenzweig V (1987) Identification of the complement decay-accelerating factor (DAF) on epithelium and glandular cells and in body fluids. J Exp Med 165: 848–864

Merry AH, Rawlinson VI, Uchikawa M, Daha MR, Sim RB (1989) Studies on the sensitivity to complement-mediated lysis of erythrocytes (Inab phenotype) with a deficiency of DAF (decay accelerating factor). Br J Haematol 73: 248–253

Moran P, Raab H, Kohr WJ, Caras IW (1991) Glycophospholipid membrane anchor attachment. J Biol Chem 266: 1250–1257

Nicholson-Weller A, Burge J, Austen KF (1981) Purification from guinea pig erythrocyte stroma of a decay-accelerating factor for the classical C3 convertase, C4b,2a. J Immunol 127: 2035–2039

Nicholson-Weller A, Burge J, Fearon DT, Weller PF, Austen KF (1982) Isolation of a human erythrocyte membrane glycoprotein with decay-accelerating activity for C3 convertases of the complement system. J Immunol 129: 184–189

Nicholson-Weller A, March JP, Rosenfeld SI, Austen KF (1983) Affected erythrocytes of patients with paroxysmal nocturnal hemoglobinuria are deficient in the complement regulatory protein, decay accelerating factor. Proc Natl Acad Sci USA 80: 5066–5070

Nicholson-Weller A, March JP, Rosen CE, Spicer DB, Austen KF (1985a) Surface membrane expression by human blood leukocytes and platelets of decay-accelerating factor, a regulatory protein of the complement system. Blood 65: 1237–1244

Nicholson-Weller A, Spicer DB, Austen KF (1985b) Deficiency of the complement regulatory protein, "decay-accelerating factor", on membranes of granulocytes, monocytes, and platelets in paroxysmal nocturnal hemoglobinuria. N Engl J Med 312: 1091–1097

Nicholson-Weller A, Russian D, Austen KF (1986) Natural killer cells are deficient in the surface expression of the complement regulatory protein, decay accelerating factor (DAF). J Immunol 137: 1275–1279

Okada H, Nagami Y, Takahashi K, Okada N, Hideshima T, Takizawa H, Kondo J (1989) 20 KDa homologous restriction factor of complement resembles T cell activating protein. Biochem Biophys Res Commun 162: 1553–1559

Opferkuch W, Loos M, Borsos T (1971) Isolation and characterization of a factor from human and guinea pig serum that accelerates the decay of the SAC142. J Immunol 107: 21–26

Pangburn MK (1986) Differences between the binding sites of the complement regulatory proteins DAF, CR1, and factor H on C3 convertases. J Immunol 136: 2216–2221

Pangburn MK, Schreiber RD, Müller-Eberhard HJ (1983) Deficiency of an erythrocyte membrane protein with complement regulatory activity in paroxysmal nocturnal hemoglobinuria. Proc Natl Acad Sci USA 80: 5430–5434

Pearce EJ, Hall BF, Sher A (1990) Host-specific evasion of the alternative complement pathway by schistosomes correlates with the presence of a phospholipase C- sensitive surface molecule resembling human decay accelerating factor. J Immunol 144: 2751–2756

Perkins SJ, Haris PI, Sim RB, Chapman D (1988) A study of the structure of human complement component factor H by Fourier transform infrared spectroscopy and secondary structure averaging methods. Biochemistry 27: 4004–4012

Post TW, Arce MA, Liszewski MK, Thompson ES, Atkinson JP, Lublin DM (1990) Structure of the gene for human complement protein decay accelerating factor. J Immunol 144: 740–744

Quigg RJ, Nicholson-Weller A, Cybulsky AV, Badalamenti J, Salant DJ (1989) Decay accelerating factor regulates complement activation on glomerular epithelial cells. J Immunol 142: 877–882

Reddy P, Caras I, Krieger M (1989) Effects of O-linked glycosylation on the cell surface expression and stability of decay-accelerating factor, a glycophospholipid-anchored membrane protein. J Biol Chem 264: 17329–17336

Reid KBM, Bentley DR, Campbell RD, Chung LP, Sim RB, Kristensen T, Tack BF (1986) Complement system proteins which interact with C3b or C4b. A superfamily of structurally related proteins. Immunol Today 7: 230–234

Rey-Campos J, Rubinstein P, Rodriguez de Cordoba S (1987) Decay-accelerating factor. Genetic polymorphism and linkage to the RCA (regulator of complement activation) gene cluster in humans. J Exp Med 166: 246–252

Rhee SG, Suh PG, Ryu SH, Lee SY (1989) Studies of inositol phospholipid-specific phospholipase C. Science 244: 546–550

Rimoldi MT, Sher A, Heiny S, Lituchy A, Hammer CH, Joiner K (1988) Developmentally regulated expression by *Trypanosoma cruzi* of molecules that accelerate the decay of complement C3 convertases. Proc Natl Acad Sci USA 85: 193–197

Roberts WN, Wilson JG, Wong W, Jenkins DJ, Fearon DT, Austen KF, Nicholson-Weller A (1985) Normal function of CR1 on affected erythrocytes of patients with paroxysmal nocturnal hemoglobinuria. J Immunol 134: 512–517

Robinson PJ (1991) Phosphatidylinositol membrane anchors and T-cell activation. Immunol Today 12: 35–41

Rosse W (1973) Variations in the red cells in paroxysmal nocturnal haemoglobinuria. Br J Haematol 24: 327–342

Schönermark S, Rauterberg EW, Shin ML, Loke S, Roelcke D, Hänsch GM (1986) Homologous species restriction in lysis of human erythrocytes: a membrane-derived protein with C8-binding capacity functions as an inhibitor. J Immunol 136: 1772–1776

Seya T, Farries T, Nickells M, Atkinson JP (1987) Additional forms of human decay-accelerating factor (DAF). J Immunol 139: 1260–1267

Shin ML, Hänsch G, Hu VW, Nicholson-Weller A (1986) Membrane factors responsible for homologous species restriction of complement-mediated lysis: evidence for a factor other than DAF operating at the stage of C8 and C9. J Immunol 136: 1777–1782

Spring FA, Judson PA, Daniels GL, Parsons SF, Mallinson G, Anstee DJ (1987) A human cell-surface glycoprotein that carries Cromer-related blood group antigens on erythrocytes and is also expressed on leucocytes and platelets. Immunology 62: 307–313

Stafford HA, Tykocinski ML, Lublin DM, Holers VM, Rosse WF, Atkinson JP, Medof ME (1988) Normal polymorphic variations and transcription of the decay accelerating factor gene in paroxysmal nocturnal hemoglobinuria cells. Proc Natl Acad Sci USA 85: 880–884

Sugita Y, Uzawa M, Tomita M (1987) Isolation of decay-accelerating factor (DAF) from rabbit erythrocyte membranes. J Immunol Methods 104: 123–130

Sugita Y, Nakano Y, Tomita M (1988) Isolation from human erythrocytes of a new membrane protein which inhibits the formation of complement transmembrane channels. J Biochem (Tokyo) 104: 633–637

Tausk F, Fey M, Gigli I (1989) Endocytosis and shedding of the decay accelerating factor on human polymorphonuclear cells. J Immunol 143: 3295–3302

Telen MJ, Green AM (1989) The Inab phenotype: characterization of the membrane protein and complement regulatory defect. Blood 74: 437–441

Telen MJ, Hall SE, Green AM, Moulds JJ, Rosse WF (1988) Identification of human erythrocyte blood group antigens on decay-accelerating factor (DAF) and an erythrocyte phenotype negative for DAF. J Exp Med 167: 1993–1998

Thomas J, Webb W, Davitz MA, Nussenzweig V (1987) Decay accelerating factor diffuses rapidly on HeLa cell surfaces. Biophys J 51: 522a

Volarevic S, Burns CM, Sussman JJ, Ashwell JD (1990) Intimate association of Thy-1 and T-cell antigen receptor with the CD45 tyrosine phosphatase. Proc Natl Acad Sci USA 87: 7085–7089

Walter EI, Roberts WL, Rosenberry TL, Ratnoff WD, Medof ME (1990) Structural basis for variations in the sensitivity of human decay accelerating factor to phosphatidylinositol-specific phospholipase C cleavage. J Immunol 144: 1030–1036 (published erratum: J Immunol 1990, 144: 4072)

Walthers L, Salem M, Tessel J, Laird-Fryer B, Moulds JJ (1983) The Inab phenotype: another example found. Transfusion 23: 423a

Yamamoto H, Blaas P, Nicholson-Weller A, Hänsch GM (1990) Homologous species restriction of the complement-mediated killing of nucleated cells. Immunology 70: 422–426

Zalman L, Wood LM, Müller-Eberhardt HJ (1986) Isolation of a human erythrocyte membrane protein capable of inhibiting expression of homologous complement transmembrane channels. Proc Natl Acad Sci USA 83: 6975–6979

Zalman L, Wood LM, Frank MM, Müller-Eberhard HJ (1987) Deficiency of the homologous restriction factor in paroxysmal nocturnal hemoglobinuria. J. Exp. Med. 165: 572–577

Complement Receptor Type 1

G. D. Ross

1	Introduction and Historical Perspective	31
1.1	Overview	31
1.2	Discovery of CR1 and Early Research	33
1.3	Isolation and Structural Analysis	34
2	Structural and Molecular Biology	34
2.1	Structural Analysis of Isolated CR1 Protein	34
2.2	Molecular Cloning of CR1 and Analysis of Its cDNA Sequence	35
3	Function of Erythrocyte CR1	36
3.1	Specificity for C3b, C4b, and iC3b	36
3.2	Inhibition of C3 and C5 Convertases	36
3.3	Factor I Cofactor Activity	37
3.4	Transport of Immune Complexes to the Macrophage Phagocytic System	38
3.5	Effect of Inheritance vs Disease Processes on the Number of CR1 per Erythrocyte	39
3.6	Immunotherapy with sCR1	40
References		41

1 Introduction and Historical Perspective

1.1 Overview

This review will focus on the complement receptor type 1 (CR1, CD35) expressed by erythrocytes and will cover its structure, molecular biology, and function as a membrane inhibitor of complement activation. The CR1 present on phagocytic cells and lymphocytes has similar functions in regulation of complement activation and serves as a receptor responsible for triggering cellular activation events such as phagocytosis or Ig synthesis. These latter receptor functions of CR1 are not covered in this review, and the reader is referred to several past reviews about leukocyte CR1 for this information (Ross and Medof 1985; Wright and Griffin 1985; Fearon and Ahearn 1989; Ross et al. 1989). Erythrocyte CR1 has three functions in regulation of complement activation that are covered in this review: (a) inhibition of the C3 and C5 convertases of the classical and

Department of Microbiology and Immunology, University of Louisville, Louisville, KY 40292, USA

Table 1. Summary of CR1[a] features

Structure	Receptor specificity	Monoclonal antibodies	Cell type distribution	Functions
Single glycoprotein chain with one of four allotypic sizes A, 250 kDa; frequency, 0.82 B, 290 kDa; frequency, 0.18 C, 330 kDa (rare) D, 210 kDa (very rare) Allotypic variation due to different numbers of LHR units of protein sequence that make up external membrane domain; each LHR consists of seven SCR motifs of ~61 amino acids that are common to all members of the RCA gene family of chromosome 1.	High affinity for C3b and lower affinity for C4b and iC3b	44D, 57F, C3To5, E11, Yz-1, 3D9, J3B11, J3D3	Erythrocytes, neutrophils, monocytes, macrophages, eosinophils, mature B cells, a small subset of T cells, follicular dendritic cells, Langerhans cells, Kupffer cells, glomerular podocytes, and peripheral neurons	Inhibits both classical and alternative pathway C3 and C4 convertases; serves as a cofactor for factor I cleavage of C3b or C4b; essential cofactor for factor I cleavage of iC3b into C3c and C3dg. Erythrocyte CR1 serves to transport immune complexes to macrophages for clearance. Phagocyte CR1 promotes the adherence and phagocytosis of particles and immune complexes. Lymphocyte CR1 enhances antigen recognition with low antigen doses.

LHR, long homologous repeat; SCR, short consensus repeat; RCA, regulator of C activation
[a] Alternative names for CR1 found in the literature are: C3b receptor, C3b/C4b receptor, immune adherence receptor, and CD35

alternative pathways of complement activation; (b) factor I cofactor activity for cleavage of C3b and iC3b; (c) adsorption of soluble immune complexes, thereby inhibiting complement-mediated inflammation. Table 1 summarizes the attributes of CR1 on all cell types.

1.2 Discovery of CR1 and Early Research

The first description of a complement receptor activity on erythrocytes was reported in a study of trypanosome immunity. DUKE and WALLACE (1930) noted that when trypanosomes were mixed with blood from patients with trypanosomiasis, the trypanosomes became thoroughly coated with the red blood cells. Fresh immune serum caused the trypanosomes to be coated with red cells from humans or monkeys but not with red cells from guinea pigs. With immune guinea pig blood, they noted that the platelets rather than the red cells became bound to trypanosomes. They showed that red cell binding to trypanosomes required both a heat stable component from immune serum and a heat labile component from normal guinea pig serum. Subsequently, WALLACE and WORMALL proposed in 1931 that complement was required for this adherence reaction (WALLACE and WORMALL 1931). By 1938, BROWN and BROOM had shown that treatment of serum with other known complement inactivators (i.e., ammonia, cobra venom) also prevented red cell binding to trypanosomes. Later, in 1953, NELSON described red cell binding to serum-opsonized bacteria and termed the phenomenon immune adherence. He observed that, in immune blood, bacteria were bound preferentially to erythrocytes rather than to leukocytes. However, after prolonged incubation of bacteria in whole blood, red cell bound bacteria were ingested by the blood leukocytes. Considerably less leukocyte ingestion was observed in leukocyte-serum mixtures devoid of red cells. Accordingly, NELSON proposed that bacteria bound to erythrocytes might be more easily ingested by phagocytic cells than were individual uncomplexed bacteria (NELSON 1953). Many years later, this hypothesis was shown to be correct and erythrocyte CR1 was shown to have the function of transporting bacteria and immune complexes to tissue macrophages. By 1963, NISHIOKA and LINSCOTT had demonstrated that immune adherence required complement activation only up to the C3 stage and postulated that erythrocytes bore immune adherence receptors specific for the fixed C3 on bacteria. By 1968 it had been demonstrated that the phagocytosis of antibody-coated sheep erythrocytes by guinea pig polymorphonuclear leukocytes also required only the first four components of complement (C1, C4, C2, and C3), and it was proposed that phagocytic cells bore immune adherence receptors that were analogous to those of erythrocytes (GIGLI and NELSON 1968). The earlier finding by DUKE and WALLACE of guinea pig platelet adherence was explained by studies reported by Peter HENSON in 1969. Guinea pig platelets, but not guinea pig erythrocytes were found to have receptors for particle bound C3 (Henson 1969). NELSON'S finding of red cell adherence to bacteria was extended to soluble immune complexes by MEDOF and coworkers in 1982. They demonstrated that soluble immune complexes in blood also bound preferentially to erythrocytes rather than to leukocytes (MEDOF and OGER 1982).

1.3 Isolation and Structural Analysis

CR1 was first isolated from erythrocytes in 1979 by FEARON in an attempt to characterize the erythrocyte membrane inhibitor of complement activation (FEARON 1979). A glycoprotein of 205 kDa was isolated by virtue of its ability both to accelerate the decay of the alternative pathway C3/C5 convertase on sheep erythrocytes (EAC3b, Bb, P) and bind specifically to C3-Sepharose in a manner resembling plasma factor H. The isolated gp205 also resembled factor H in its ability to serve as a cofactor for factor I in the cleavage of C3b into iC3b. Subsequently, a rabbit antibody raised to the isolated gp205 was shown to block the C3b receptor activity of lymphocytes, monocytes, and neutrophils, leading to the conclusion that gp205 was probably CR1 (FEARON 1980). Others confirmed these findings and showed that rabbit anti-gp205 inhibited CR1-dependent C4b receptor activity but had no effect on CR2-dependent C3d receptor activity (DOBSON et al. 1981). Sequence analysis of tryptic peptides of CR1 permitted the generation of oligonucleotide probes and the cloning of CR1 cDNA for use in complete analysis of protein and gene structure (KLICKSTEIN et al. 1987; WONG et al. 1985, 1989).

2 Structural and Molecular Biology

2.1 Structural Analysis of Isolated CR1 Protein

Although initial investigations had suggested that CR1 consisted of a single 205 kDa glycoprotein, later studies found some variation in the size of CR1 isolated from different individuals. Four allotypes of CR1 have been characterized that vary widely in molecular mass. The most common A or F allotype, with a gene frequency of 0.82, has a molecular mass of 190 kDa when analyzed by nonreducing SDS-PAGE and 250 kDa when analyzed following reduction of disulfide bonds. The B or S allotype, with a gene frequency of 0.18, appears to be 220 kDa when analyzed under nonreducing conditions or 290 kDa after disulfide bond reduction (DYKMAN et al. 1983; WONG et al. 1983). Rare CR1 allotypes of higher (C) or lower (D) molecular mass have also been observed (Dykman et al. 1984, 1985). Cells from a sinlge individual can be homozygous and express only a single CR1 allotype or can be heterozygous and express two distinct CR1 allotypes. The basis for the large difference in molecular mass between the allotypes was shown to be a protein difference rather than a difference in glycosylation, and later genetic studies showed that it was probably due to duplication of homologous domains.

2.2 Molecular Cloning of CR1 and Analysis of Its cDNA Sequence

CR1 was the first of the complement receptors to be cloned and seqenced. Unique oligonucleotide probes were used to identify a partial cDNA clone of CR1. Later, full-length cDNA clones were also produced (WONG et al. 1985; KLICKSTEIN et al. 1987). The external membrane domain of CR1 consists of a series of repeating sequence motifs of ~ 61 amino acids that are known as short consensus repeats (SCRs). A large family of proteins contains similar SCR sequences, and most of these are members of the complement system whose function involves regulation of complement activation through binding to C3b and C4b. Members of this gene family are encoded by genes mapping to chromosome 1, band q32, in a region now called the regulator of complement activation (RCA) (WEIS et al. 1987; CARROLL et al. 1988; REY-CAMPOS et al. 1988; LUBLIN et al. 1988). Other gene family members of the RCA include: CR2, factor H, C4 binding protein (C4bp), membrane cofactor protein (MCP), and decay accelerating factor (DAF). The gene for CR2 is closely linked to the CR1 gene, and in the mouse it has been shown that a single gene encodes both CR1 and CR2 via alternative splicing of mRNA (SEYA et al. 1990).

It has been proposed that each SCR forms a protein loop held together by disulfide bonds derived from four conserved cysteine residues (JANATOVA et al. 1989). In the A allotype of CR1, 28 of the 30 SCR sequences are arranged in four similar groups of seven SCRs. Each of these repeating groups of seven SCRs form a long homologous repeat (LHR) unit. In the B allotype of CR1, there is an additional fifth LHR that may have been derived from a homologous crossover event involving the terminal LHR of the A allotype (WONG et al. 1989). The LHRs of the A allotype have been designated A, B, C, and D types, whereas the fifth LHR of the B allotype is designated A/B because of its sequence similarity to both the A and B LHR. The C3b and C4b binding domains of CR1 have been localized to the terminal three LHR (A, B, and C) of the A allotype. These studies were facilitated by the positioning of convenient restriction sites at the boundaries of each LHR. This allowed recombinant CR1 to be investigated by selectively deleting one or more of the LHRs (KLICKSTEIN et al. 1988).

These repeating LHR units of homologous protein structure give rise to several repeating antigenic epitopes as demonstrated by studies with monoclonal antibodies (mAbs) after deletion mutagenesis (KLICKSTEIN et al. 1988). In addition, studies with soluble recombinant CR1 (sCR1) have demonstrated that CR1 is able to bind C3b and/or C4b bivalently, indicating that the two binding sites demonstrated by deletion mutagenesis with recombinant membrane CR1 are able to function simultaneously (WEISMAN et al. 1990). This ability to bind homodimers of C3b or C4b or C3b/C4b heterodimers appears to be important in CR1's function in dissociating the C3 and C5 convertases of the classical and alternative pathways of complement activation.

3 Function of Erythrocyte CR1

3.1 Specificity for C3b, C4b, and iC3b

NELSON and coworkers, who first showed that C3 was required for immune adherence, later noted that the ability of the fixed C3 (C3b) to participate in immune adherence or activation of the terminal complement components for hemolysis was destroyed by a normal serum enzyme that they called C3 inactivator (TAMURA and NELSON 1967). Subsequently, C3 inactivator was named factor I and it was shown to function by proteolytically cleaving C3b into iC3b (PANGBURN et al. 1977). On the basis of these findings it was believed for several years that CR1 reacted only with C3b and not with iC3b. Later, however, weak binding of CR1 to fixed iC3b was demonstrated (ROSS et al. 1983), and it was shown that this activity was required for CR1 to express its factor I cofactor activity for cleavage of iC3b into C3c and C3dg (ROSS et al. 1982; MEDOF et al. 1982; MEDICUS et al. 1983). Binding of CR1 to fixed C4b was suggested by a study reported by COOPER in 1969 in which it was shown that sheep erythrocyte intermediate complexes, prepared with very large amounts of C4 (EAC1, 4b), mediated immune adherence without need for further addition of C2 and C3. Subsequently, both fluid phase C3b (ROSS and POLLEY 1975) and anti-CR1 (DOBSON et al. 1981) were shown to block C4b-dependent immune adherence. Isolated membrane CR1 (MEDOF and NUSSENZWEIG 1984) and recombinant CR1 (WEISMAN et al. 1990) have also been shown to serve as factor I cofactors for cleavage of C4b, providing further evidence for CR1 binding to C4b.

Several types of data have localized the CR1 binding site in C3 to the C3c region of C3b and iC3b. First, fluid C3c was shown to inhibit both C3b- and C4b-dependent rosettes with CR1-bearing cells (ROSS and POLLEY 1975). Later, C3c-coated fluorescent microspheres were shown to bind to CR1 (ROSS and LAMBRIS 1982). Finally, using synthetic peptides representing small portions of the C3c region of C3, a small peptide from the NH_2 terminus of the α chain of C3c was identified that bound to CR1 and appears to represent the CR1 binding site in C3 (BECHERER and LAMBRIS 1988).

3.2 Inhibition of C3 and C5 Convertases

The first complement regulatory function described for CR1 was its ability to dissociate the C5 convertase of the alternative pathway formed on sheep erythrocytes (EC3b, Bb, P) (FEARON 1979). Isolated membrane CR1 was initially shown to have this activity, and similar findings were made later with recombinant sCR1 (WEISMAN et al. 1990). CR1 binds to fixed C3b, displacing Bb and preventing more factor B from binding to C3b. Likewise, CR1 binds to fixed C4b, dissociating any C2 that may be complexed with C4b and preventing more C2 from binding to the C4b. These activities of binding to fixed C4b or C3b confer on CR1 the ability to regulate formation of the classical and alternative pathway

C3 convertases. Since CR1 has a higher affinity for C3b than for C4b, it more efficiently inhibits the alternative pathway C3 convertase than the classical pathway enzyme. Factor H, C4bp, and DAF function in an analogous manner, except that factor H does not bind to C4b and C4bp functions primarily with C4b rather than with C3b. Although membrane DAF has been shown to provide some degree of protection to host cells from C3 or C5 convertase, it has not been possible to demonstrate the binding of DAF to C3b or C4b. It has been claimed that CR1 differs from factor H in that CR1 is not restricted from binding to C3b on surfaces that activate the alternative pathway (FEARON 1979; WEISMAN et al. 1990). However, this has not been rigorously demonstrated by binding affinity studies using sCR1 in the same way as was reported earlier with factor H (KAZATCHKINE et al. 1979). Moreover, other studies have noted that factor H-restricted C3b on rabbit erythrocytes bound poorly to neutrophil CR1 as compared to factor H-sensitive C3b on sheep erythrocytes (G.D. Ross, unpublished observation).

Despite its ability to promote the decay of C3 and C5 convertases, CR1 does not appear to play an essential role in the protection of erythrocytes from homologous C attack. Anti-CR1 does not increase the complement sensitivity of red cells. By contrast, blockade or removal of DAF is associated with increased red cell sensitivity to complement, suggesting that DAF, rather than CR1, is the essential regulator of complement activation on red blood cells (MEDOF et al. 1984, 1987). It is unknown whether CR1 may play a more important role in regulation of complement activation on cells other than erythrocytes that may not express DAF, such as peripheral nerve cells that have recently been shown to express CR1 (VEDELER and MATRE 1988; VEDELER et al. 1989, 1990).

3.3 Factor I Cofactor Activity

Early investigations of isolated CR1 demonstrated its factor H-like cofactor activity in mediating the cleavage of fluid phase C3b by factor I (FEARON 1979). On a weight basis, CR1 is more effective than is factor H and, unlike factor H, is also able to function efficiently as a cofactor for factor I cleavage of C4b (MEDOF and NUSSENZWEIG 1984). However, unlike all other factor I cofactors, CR1 is also able to function efficiently as a cofactor for cleavage of iC3b into C3dg and C3c (Ross et al. 1982; MEDOF et al. 1982; MEDICUS et al. 1983). This was demonstrated first with fixed iC3b (Ross et al. 1982) and then was subsequently confirmed with fluid phase C3b and C4b-like C4ma (methylamine inactivated C4) using soluble recombinant CR1 (WEISMAN et al. 1990).

Despite the greater activity of soluble CR1 in mediating factor I cleavage of C3b into iC3b, factor H is the major cofactor in blood used for this step in C3b metabolism. There are only small amounts of CR1 in the blood and all of it is membrane bound. By contrast, factor H functions very inefficiently as a cofactor in the breakdown of iC3b into C3dg and C3c. Prolonged incubation of factors H and I with iC3b is required for cleavage unless low ionic strength conditions are employed (Ross et al. 1982; JANATOVA and GOBEL 1985). The binding affinity of factor H for fixed iC3b is very low and was only measurable in low ionic strength buffer (Ross et al. 1983), probably indicating why factor H cofactor activity for the iC3b substrate is difficult to detect in isotonic buffer. Although CR1 resembles factor H in exhibiting reduced binding to fixed iC3b as compared to fixed C3b, CR1 binding to fixed iC3b is readily detectable in isotonic salt buffer, allowing CR1 to have much greater cofactor activity than factor H under physiologic conditions.

The ability of CR1 to promote the breakdown of fixed iC3b may be one of the most important functions of CR1 in regulating complement injury of host tissue. iC3b is deposited in normal tissue by the incomplete clearance of immune complexes and by some types of vascular injury such as cardiac ischemia (CRAWFORD et al. 1988). The fixed iC3b in tissues mediates the attachment and degranulation of neutrophils via membrane iC3b receptors (CR3) (Ross et al. 1989; MARKS et al. 1989). Resulting inflammatory damage can be reduced by depleting total blood complement activity with cobra venom factor or by blocking neutrophil CR3 with anti-CR3 (CRAWFORD et al. 1988; NOURSHARGH et al. 1989; SIMPSON et al. 1990). Endogenous membrane CR1 may also provide some protective effect to host cells in either inhibiting further complement activation or accelerating iC3b cleavage. This has been demonstrated with lymphocyte CR1 (IIDA and NUSSENZWEIG 1981, 1983) and may also occur with other cell types such as kidney podocytes (FISCHER et al. 1986) or peripheral nerve cells (VEDELER and MATRE 1990).

3.4 Transport of Immune Complexes to the Macrophage Phagocytic System

Erythrocytes use CR1 to adsorb both soluble and particulate immune complexes, preventing their deposition in blood vessels where they could otherwise mediate inflammation and tissue damage. Due to the greater abundance of red cells as compared to leukocytes, immune complexes that enter the blood bind primarily to red cells and not to leukocytes (MEDOF and OGER 1982). Despite expressing only small numbers of CR1 per cell, erythrocytes are able to bind immune complexes efficiently because their CR1 are primarily located in large clusters on the red cell surface (CHEVALIER and KAZATCHKINE 1989; PACCAUD et al. 1990). Red cells carry C3-coated immune complexes via their membrane CR1 to the liver where macrophages strip the complexes from the red cell surface and return the red cells to the circulation (CORNACOFF et al. 1983; SCHIFFERLI et al. 1986, 1988). The process is rapid and effective, even in patients with systemic lupus erythematosus (SLE), in whom both plasma complement and erythrocyte CR1 have been depleted by chronic exposure to immune complexes (SCHIFFERLI et al. 1989).

3.5 Effect of Inheritance vs Disease Processes on the Number of CR1 per Erythrocyte

In addition to an allotypic variation in the size of CR1, the number of CR1 per erythrocyte (CR1/E) is also a heritable trait (WILSON et al. 1982). A restriction fragment length polymorphism (RFLP) that correlates with either low or high numbers of CR1/E has been identified in the CR1 gene using Hind III (WILSON et al. 1986). Individuals may express either very low numbers of CR1/E (6.9 kb Hind III fragment), high numbers of CR1/E (7.4 kb fragment), or have intermediate numbers of CR1/E (heterozygotes with both 6.9 and 7.4 kb fragments). There is a relatively broad range of CR1/E values associated with each phenotype and some overlapping ranges of CR1/E values among the three allotypes. Thus, analysis of the CR1/E value does not provide a clear indication of an individual's CR1/E genotype. Most individuals are either homozygous high or heterozygotes, as the frequency of the 6.9 kb allotype is very low.

CR1/E values have been measured using either ^{125}I-C3b dimers (WILSON et al. 1982) or ^{125}I-anti-CR1 mAbs (IIDA et al. 1982; WALPORT et al. 1985; ROSS et al. 1985; THOMSEN et al. 1987; YEN et al. 1989; TAUSK et al. 1990). A somewhat lower mean value for normal individuals was obtained with C3b dimers (350–450 CR1/E) than with anti-CR1 mAbs (550–750 CR1/E). This may have been due to the presence of repeating antigenic epitopes in CR1 that are expressed in adjacent LHR units (BARTOW et al. 1989). However, studies with recombinant CR1 have also indicated that individual CR1 molecules can bind two C3b dimers (WEISMAN et al. 1990), so neither method is probably ideal for quantitation of CR1/E.

Initial tests of patients with SLE suggested that both the patients and their first degree relatives had low CR1/E values ($\leq 50\%$ of normal volunteer mean), and it was proposed that inheritance of low numbers of CR1/E might be a risk factor that contributed to the onset of this autoimmune disease, whose pathogenesis is associated with incomplete clearance of circulating autoantibody immune complexes (WILSON et al. 1982). Although several laboratories confirmed the finding of low CR1/E in patients with SLE, some anomalies of CR1/E expression were noted in other family studies that could not be explained solely by inheritance (WALPORT et al. 1985). Moreover, a strong correlation was established between the presence of low CR1/E and evidence of C3 fragment (C3dg) deposition on the patients' erythrocytes (ROSS et al. 1985). This finding of a correlation between low CR1/E and elevated C3dg/E was extended to several other diseases, all involving complement activation with or without the presence of circulating immune complexes (ROSS et al. 1985; TAUSK et al. 1986; COHEN et al. 1988). Some patients tested serially over the course of their disease showed increased CR1/E during periods when C3 deposition on their erythrocytes was reduced and, conversely, exhibited diminished CR1/E during periods of apparent active disease when there was an increase in the number of C3 fragments per erythrocyte. When patients with immune complex-associated disease were transfused with erythrocytes bearing large numbers of CR1/E, the transfused erythrocytes soon acquired C3 fragments and lost a proportion of their CR1

(WALPORT et al. 1987). In vitro studies with red cells and macrophages also suggested that macrophages might remove CR1 in the process of removing immune complexes from the red cell surface (DAVIES et al. 1990). Chronic infusion of immune complexes into primates also resulted in lower CR1/E (COSIO et al. 1990). Finally, studies of patients with SLE and their families for the 6.9 kb RFLP associated with low CR1/E showed no correlation of the RFLP with SLE (MOLDENHAUER et al. 1987). Even families with a history of SLE did not exhibit an abnormally higher incidence of the low CR1/E RFLP (TEBIB et al. 1989).

Despite the lack of association of low CR1/E inheritance with increased risk for SLE, available evidence suggests that erythrocyte CR1 may play an important role in regulating complement activation and promoting the clearance of immune complexes. Even in patients with both low CR1/E and low total plasma complement activity, erythrocytes mediated the rapid clearance of experimentally infused immune complexes (SCHIFFERLI et al. 1988, 1989). Thus, individuals with the low CR1/E phenotype are capable of efficient clearance of immune complexes, and this may be why they are not at increased risk for developing SLE.

Even though low CR1/E inheritance does not appear to increase the risk of SLE, the process of immune complex transport and removal via erythrocyte CR1 appears to be important in regulating pathogenic levels of circulating immune complexes. Investigations of patients who have only one of two functional C4 genes (C4A or C4B) have shown that an inherited deficiency of the C4A gene product increases the statistical risk of acquiring SLE (FIELDER et al. 1983; DUNCKLEY et al. 1987). The C4A protein functions more efficiently than the C4B protein in opsonizing immune complexes for transport by erythrocyte CR1. Accordingly, it has been proposed that the critical function of C4A that is missing in patients who have only C4B is the ability to opsonize immune complexes and adsorb them to erythrocytes via CR1 (GATENBY et al. 1990).

3.6 Immunotherapy with sCR1

The usefulness of CR1 as a regulator of complement activation is limited by both its low concentration in blood and localization to erythrocyte membranes. Fearon and coworkers (WEISMAN et al. 1990) have recently reported the development of a sCR1 that was constructed without the transmembrane region and cytoplasmic tail. Large amounts of this sCR1 were generated in transfected Chinese hamster ovary cells grown in a bioreactor. The sCR1 was then purified to homogeneity with preparative cation exchange FPLC (YEH et al. 1991). This sCR1 would be expected to: (a) reduce C3b deposition by inhibiting C3 convertase, (b) reduce both C5a-mediated inflammation and damage from the membrane attack complex (C5b-9) by inhibiting C5 convertase, and (c) accelerate the breakdown of fixed iC3b into fixed C3dg through its factor I cofactor activity. Two in vivo investigations with a rat model demonstrated the efficacy of sCR1 therapy for reduction of complement-induced tissue injury. First, sCR1 was shown to reduce infract size in a model of postischemic reperfusion injury (WEISMAN et al. 1990). The mechanism for complement activation in this model is unknown but may be due to activation of the alternative pathway by the damaged blood vessel wall. The sCR1 was also very effective in reducing inflammation and tissue injury following intradermal injection of immune complexes in a rat model of the passive reversed Arthus reaction (YEH et al. 1991). Even better regulation of complement and neutrophil-mediated tissue injury would be expected in humans as compared to rats, as sCR1 is less effective in inhibiting rat than human complement. Thus, sCR1 could prove to be a valuable therapeutic agent for the treatment of all types of autoimmune and inflammatory diseases that involve complement activation.

References

Bartow TJ, Klickstein LB, Fearon DT (1989) Localization of monoclonal antibody epitopes on CR1 by deletion mutagenesis. Compl Inflamm 6: 312 (abstract)

Becherer JD, Lambris JD (1988) Identification of the C3b receptor-binding domain in third component of complement. J Biol Chem 263: 14586–14591

Brown HC, Broom JC (1938) Studies in trypanosomiasis. II. Observations on the red cell adhesion test. Trans R Soc Trop Med Hyg 32: 209–222

Carroll MC, Alicot EM, Katzman PJ, Klickstein LB, Smith JA, Fearon DT (1988) Organization of the genes encoding complement receptors type 1 and 2, decay-accelerating factor, and C4-binding protein in the RCA locus on human chromosome 1. J Exp Med 167: 1271–1280

Chevalier J, Kazatchkine MD (1989) Distribution in clusters of complement receptor type one (CR1) on human erythrocytes. J Immunol 142: 2031–2036

Cohen JHM, Autran B, Jouvin MH, Aubry JP, Rozenbaum W, Banchereau J, Debre P, Revillard JP, Kazatchkine M (1988) Decreased expression of the receptor for the C3b fragment of complement on erythrocytes of patients with acquired immunodeficiency syndrome. Presse Med 17: 727–730

Cooper NR (1969) Immune adherence by the fourth component of complement. Science 165: 396–398

Cornacoff JB, Hebert LA, Smead WL, VanAman ME, Birmingham DJ, Waxman FJ (1983) Primate erythrocyte-immune complex-clearing mechanism. J Clin Invest 71: 236–247

Cosio FG, Xiao-Ping S, Birmingham DJ, Aman MV, Hebert LA (1990) Evaluation of the mechanisms responsible for the reduction in erythrocyte complement receptors when immune complexes form in vivo in primates. J Immunol 145: 4198–4206

Crawford MH, Grover FL, Kolb WP, McMahan CA, O'Rourke RA, McManus LM, Pinckard RN (1988) Complement and neutrophil activation in the pathogenesis of ischemic myocardial injury. Circulation 78: 1449–1458

Davies KA, Hird V, Stewart S, Sivolapenko GB, Jose P, Epenetos AA, Walport MJ (1990) A study of in vivo immune complex formation and clearance in man. J Immunol 144: 4613–4620

Dobson NJ, Lambris JD, Ross GD (1981) Characteristics of isolated erythrocyte complement receptor type one (CR1, C4b-C3b receptor) and CR1-specific antibodies. J Immunol 126: 693–698

Duke HL, Wallace JM (1930) "Red-cell adhesion" in trypanosomiasis of man and animals. Parasitology 22: 414–456

Dunckley H, Gatenby PA, Hawkins B, Naito S, Serjeantson SW (1987) Deficiency of C4A is a genetic determinant of systemic lupus erythematosus in three ethnic groups. J Immunogenet 14: 209–218

Dykman TR, Cole JL, Iida K, Atkinson JP (1983) Polymorphism of human erythrocyte C3b/C4b receptor. Proc Natl Acad Sci USA 80: 1698–1702

Dykman TR, Hatch JA, Atkinson JP (1984) Polymorphism of the human C3b/C4b receptor. Identification of a third allele and analysis of receptor phenotypes in families and patients with systemic lupus erythematosus. J Exp Med 159: 691–703

Dykman TR, Hatch JA, Aqua MS, Atkinson JP (1985) Polymorphism of the C3b/C4b receptor (CR1): characterization of a fourth allele. J Immunol 134: 1787–1789

Fearon DT (1979) Regulation of the amplification C3 convertase of human complement by an inhibitory protein isolated from human erythrocyte membrane. Proc Natl Acad Sci USA 76: 5867–5871

Fearon DT (1980) Identification of the membrane glycoprotein that is the C3b receptor of the human erythrocyte, polymorphonuclear leukocyte, B lymphocyte, and monocyte. J Exp Med 152: 20–30

Fearon DT, Ahearn JM (1989) Complement receptor type 1 (C3b/C4b receptor, CD35) and complement receptor type 2 (C3d/Epstein-Barr virus receptor; CD21). In: Lambris JD (ed) The third component of complement. Chemistry and biology. Springer, Berlin Heidelberg New York, pp 83–98 (Current topics in microbiology and immunology, vol 153)

Fielder AHL, Walport MJ, Batchelor JR, Rynes RI, Black CM, Dodi IA, Hughes GRV (1983) Family study of the major histocompatibility complex in patients with systemic lupus erythematosus: importance of null alleles of C4A and C4B in determining disease susceptibility. Br Med J 286: 425–428

Fischer E, Appay MD, Cook J, Kazatchkine MD (1986) Characterization of the human glomerular C3 receptor as the C3b/C4b complement type one (CR1) receptor. J Immunol 136: 1373–1377

Gatenby PA, Barbosa JE, Lachmann PJ (1990) Differences between C4A and C4B in the handling of immune complexes: the enhancement of CR1 binding is more important than the inhibition of immunoprecipitation. Clin Exp Immunol 79: 158–163

Gigli I, Nelson RA Jr (1968) Complement dependent immune phagocytosis. I. Requirements for C'1, C'2, C'3, C'4. Exp Cell Res 51: 45–67

Henson PM (1969) The adherence of leucocytes and platelets induced by fixed IgG antibody or complement. Immunology 16: 107–121

Iida K, Nussenzweig V (1981) Complement receptor is an inhibitor of the complement cascade. J Exp Med 153: 1138–1150

Iida K, Nussenzweig V (1983) Functional properties of membrane-associated complement receptor CR1. J Immunol 130: 1876–1880

Iida K, Mornaghi R, Nussenzweig V (1982) Complement receptor (CR1) deficiency in erythrocytes from patients with systemic lupus erythematosus. J Exp Med 155: 1427–1438

Janatova J, Gobel RJ (1985) Activation and fragmentation of the third (C3) and the fourth (C4) components of complement: generation and isolation of physiologically relevant fragments C3c and C4c. J Immunol Methods 85: 17–26

Janatova J, Reid KBM, Willis AC (1989) Disulfide bonds are localized within the short consensus repeat units of complement regulatory proteins: C4b-binding protein. Biochemistry 28: 4754–4761

Kazatchkine MD, Fearon DT, Austen KF (1979) Human alternative complement pathway: membrane-associated sialic acid regulates the competition between B and β1H for cell-bound C3b. J Immunol 122: 75–80

Klickstein LB, Wong WW, Smith JA, Weis JH, Wilson JG, Fearon DT (1987) Human C3b/C4b receptor (CR1). Demonstration of long homologous repeating domains that are composed of the short consensus repeats characteristic of C3/C4 binding proteins. J Exp Med 165: 1095–1112

Klickstein LB, Bartow TJ, Miletic V, Rabson LD, Smith JA, Fearon DT (1988) Identification of distinct C3b and C4b recognition sites in the human C3b/C4b receptor (CR1, CD35) by deletion mutagenesis. J Exp Med 168: 1699–1717

Lublin DM, Liszewski MK, Post TW, Arce MA, Le Beau MM, Rebentisch MB, Lemons RS, Seya T, Atkinson JP (1988) Molecular cloning and chromosomal localization of human membrane cofactor protein (MCP). J Exp Med 168: 181–194

Marks RM, Todd RF III, Ward PA (1989) Rapid induction of neutrophil-endothelial adhesion by endothelial complement fixation. Nature 339: 314–317

Medicus RG, Melamed J, Arnaout MA (1983) Role of human factor I and C3b receptor in the cleavage of surface-bound C3b. Eur J Immunol 13: 465–470

Medof ME, Nussenzweig V (1984) Control of the function of substrate-bound C4b-C3b by the complement receptor CR1. J Exp Med 159: 1669–1685

Medof ME, Oger JJ-F (1982) Competition for immune complexes by red cells in human blood. J Clin Lab Immunol 7: 7–13

Medof ME, Iida K, Mold C, Nussenzweig V (1982) Unique role of the complement receptor CR1 in the degradation of C3b associated with immune complexes. J Exp Med 156: 1739–1754

Medof ME, Kinoshita T, Nussenzweig V (1984) Inhibition of complement activation on the surface of cells after incorporation of decay-accelerating factor (DAF) into their membranes. J Exp Med 160: 1558–1578

Medof ME, Gottlieb A, Kinoshita T, Hall S, Silber R, Nussenzweig V, Rosse WF (1987) Relationship between decay accelerating factor deficiency, diminished acetylcholinesterase activity, and defective terminal complement pathway restriction in paroxysmal nocturnal hemoglobinuria erythrocytes. J Clin Invest 80: 165–174

Moldenhauer F, David J, Fielder AHL, Lachmann PJ, Walport MJ (1987) Inherited deficiency of erythrocyte complement receptor type 1 does not cause susceptibility to systemic lupus erythematosus. Arthritis Rheum 30: 961–966

Nelson RA Jr (1953) The immune adherence phenomenon: an immunologically specific reaction between micro-organisms and erythrocytes leading to enhanced phagocytosis. Science 118: 733–737

Nishioka K, Linscott WD (1963) Components of guinia pig complement. I. Separation of a serum fraction essential for immune hemolysis and immune adherence. J Exp Med 118: 767–793

Nourshargh S, Rampart M, Hellewell PG, Jose PJ, Harlan JM, Edwards AJ, Williams TJ (1989) Accumulation of ^{111}In-neutrophils in rabbit skin in allergic and non-allergic inflammatory reactions in vivo. Inhibition by neutrophil pretreatment in vitro with a monoclonal antibody recognizing the CD18 antigen. J Immunol 142: 3193–3198

Paccaud J-P, Carpentier J-L, Schifferli JA (1990) Difference in the clustering of complement receptor type 1 (CR1) on polymorphonuclear leukocytes and erythrocytes: effect on immune adherence. Eur J Immunol 20: 283–289

Pangburn MK, Schreiber RD, Müller-Eberhard HJ (1977) Human complement C3b inactivator: isolation, characterization, and demonstration of an absolute requirement for the serum protein β1H for cleavage of C3b and C4b in solution. J Exp Med 146: 257–270

Rey-Campos J, Rubinstein P, Rodriguez de Cordoba S (1988) A physical map of the human regulator of complement activation gene cluster linking the complement genes CR1, CR2, DAF, and C4BP J Exp Med 167: 664–669

Ross GD, Lambris JD (1982) Identification of a C3bi-specific membrane complement receptor that is expressed on lymphocytes, monocytes, neutrophils, and erythrocytes. J Exp Med 155: 96–110

Ross GD, Medof ME (1985) Membrane complement receptors specific for bound fragments of C3. Adv Immunol 37: 217–267

Ross GD, Polley MJ (1975) Specificity of human lymphocyte complement receptors. J Exp Med 141: 1163–1180

Ross GD, Lambris JD, Cain JA, Newman SL (1982) Generation of three different fragments of bound C3 with purified factor I or serum. I. Requirements for factor H vs CR1 cofactor activity. J Immunol 129: 2051–2060

Ross GD, Newman SL, Lambris JD, Devery-Pocius JE, Cain JA, Lachmann PJ (1983) Generation of three different fragments of bound C3 with purified factor I or serum. II. Location of binding sites in the C3 fragments for factors B and H, complement receptors, and bovine conglutinin. J Exp Med 158: 334–352

Ross GD, Yount WJ, Walport MJ, Winfield JB, Parker CJ, Fuller CR, Taylor RP, Myones BL, Lachmann PJ (1985) Disease-associated loss of erythrocyte complement receptors (CR1, C3b-receptors) in patients with systemic lupus erythematosus and other diseases involving autoantibodies and/or complement activation. J Immunol 135: 2005–2014

Ross GD, Walport MJ, Hogg N (1989) Receptors for IgG Fc and fixed C3. In: Asherson GL, Zembala M (eds) Human monocytes. Academic, London, pp 123–139

Schifferli JA, Ng YC, Peters DK (1986) The role of complement and its receptor in the elimination of immune complexes. N Engl J Med 315: 488–495

Schifferli JA, Ng YC, Estreicher J, Walport MJ (1988) The clearance of tetanus toxoid/anti-tetanus toxoid immune complexes from the circulation of humans. Complement- and erythrocyte complement receptor 1-dependent mechanisms. J immunol 140: 899–904

Schifferli JA, Ng YC, Paccaud J-P, Walport MJ (1989) The role of hypocomplementaemia and low erythrocyte complement receptor type 1 numbers in determining abnormal immune complex clearance in humans. Clin Exp Immunol 75: 329–335

Seya T, Hara T, Matsumoto M, Sugita Y, Akedo H (1990) Complement-mediated tumor cell damage induced by antibodies against membrane cofactor protein (MCP, CD46). J Exp Med 172: 1673–1680

Simpson PJ, Todd RF III, Mickelson JK, Fantone JC, Gallagher KP, Lee KA, Tamura Y, Cronin M, Lucchesi BR (1990) Sustained limitation of myocardial reperfusion injury by a monoclonal antibody that alters leukocyte function. Circulation 81: 226–237

Tamura N, Nelson RA Jr (1967) Three naturally occurring inhibitors of components of complement in guinea pig and rabbit serum. J Immunol 99: 582–589

Tausk F, Harpster E, Gigli I (1990) The expression of C3b receptors in the differentiation of discoid lupus erythematosus and systemic lupus erythematosus. Arthiritis Rheum 33: 888–892

Tausk FA, McCutchan JA, Spechko P, Schreiber RD, Gigli I (1986) Altered erythrocyte C3b receptor expression, immune complexes, and complement activation in homosexual men in varying risk groups for acquired immune deficiency syndrome. J Clin Invest 78: 977–982

Tebib JG, Martinez C, Granados J, Alarcon-Segovia D, Schur PH (1989) The frequency of complement receptor type 1 (CR1) gene polymorphisms in nine families with multiple cases of systemic lupus erythematosus. Arthritis Rheum 32: 1465–1469

Thomsen BS, Nielsen H, Andersen V (1987) Erythrocyte CR1 (C3b/C4b receptor) levels and disease activity in patients with SLE. Scand J Rheumatol 16: 339–346

Vedeler CA, Matre R (1988) Complement receptor CR1 on human peripheral nerve fibres. J Neuroimmunol 17: 315–322

Vedeler CA, Matre R (1990) Peripheral nerve CR1 express in situ cofactor activity for degradation of C3b. J Neuroimmunol 26: 51–56

Vedeler CA, Matre R, Fischer E (1989) Isolation and characterization of complement receptors CR1 from human peripheral nerves. J Neuroimmunol 23: 215–221

Vedeler CA, Scarpini E, Beretta S, Doronzo R, Matre R (1990) The ontogenesis of Fcγ receptors and complement receptors CR1 in human peripheral nerve. Acta Neuropathol (Berl) 80: 35–40

Wallace JM, Wormall A (1931) Red-cell adhesion in trypanosomiasis of man and other animals; some experiments on mechanism of reaction. Parasitology 23: 346–359

Walport MJ, Ross GD, Mackworth-Young C, Watson JV, Hogg N, Lachmann PJ (1985) Family studies of erythrocyte complement receptor type 1 levels: reduced levels in patients with SLE are acquired, not inherited. Clin Exp Immunol 59: 547–554

Walport MJ, Ng YC, Lachmann PJ (1987) Erythrocytes transfused into patients with SLE and haemolytic anaemia lose complement receptor type 1 from their cell surface. Clin Exp Immunol 69: 501–507

Weis JH, Morton CC, Bruns GAP, Weis JJ, Klickstein LB, Wong WW, Fearon DT (1987) A complement receptor locus: genes encoding C3b/C4 receptor and C3d/Epstein–Barr virus receptor map to 1q32. J Immunol 138: 312–315

Weisman HF, Bartow T, Leppo MK, Marsh HC Jr, Carson GR, Concino MF, Boyle MP, Roux KH, Weisfeldt ML, Fearon DT (1990) Soluble human complement receptor type 1: in vivo inhibitor of complement suppressing post-ischemic myocardial inflammation and necrosis. Science 249: 146–151

Wilson JG, Wong WW, Schur PH, Fearon DT (1982) Mode of inheritance of decreased C3b receptors on erythrocytes of patients with systemic lupus erythematosus. N Engl J Med 307: 981–986

Wilson JG, Murphy EE, Wong WW, Klickstein LB, Weis JH, Fearon DT (1986) Identification of a restriction fragment length polymorphism by a CR1 cDNA that correlates with the number of CR1 on erythrocytes. J Exp Med 164: 50–59

Wong WW, Wilson JG, Fearon DT (1983) Genetic regulation of a structural polymorphism of human C3b receptor. J Clin Invest 72: 685–693

Wong WW, Klickstein LB, Smith JA, Weis JH, Fearon DT (1985) Identification of a partial cDNA clone for the human receptor for complement fragments C3b/C4b. Proc Natl Acad Sci USA 82: 7711–7715

Wong WW, Cahill JM, Rosen MD, Kennedy CA, Bonaccio ET, Morris MJ, Wilson JG, Klickstein LB, Fearon DT (1989) Structure of the human CR1 gene. Molecular basis of the structural and quantitative polymorphisms and identification of a new CR1-like allele. J Exp Med 169: 847–863

Wright SD, Griffin FM Jr (1985) Activation of phagocytic cells' C3 receptors for phagocytosis. J Leukocyte Biol 38: 327–339

Yeh CG, Marsh HC Jr, Carson GR, Berman L, Concino MF, Scesney SM, Kuestner RE, Skibbens R, Donahue KA, Ip SH (1991) Recombinant soluble human complement receptor type 1 inhibits inflammation in the reversed passive arthus reaction in rats. J Immunol 146: 250–256

Yen J-H, Liu H-W, Lin S-F, Chen J-R, Chen T-P (1989) Erythrocyte complement receptor type 1 in patients with systemic lupus erythematosus. J Rheumatol 16: 1320–1325

Membrane Cofactor Protein

M. K. LISZEWSKI and J. P. ATKINSON

1	Introduction	45
2	Function	46
3	Structure	50
3.1	Phenotypic Patterns	50
3.2	Biosynthesis	50
4	Sequence Analysis and Genomic Organization	51
4.1	cDNA Isoforms	51
4.2	Genomic Organization	53
5	Correlation of Structure and Function	54
6	Discovery and Distribution	55
7	MCP of Other Species	56
8	Summary	57
	References	57

1 Introduction

Membrane cofactor protein (MCP, CD46) is a widely distributed regulatory protein that inhibits complement activation on host cells. Except for erythrocytes, it has been found on every cell examined (LISZEWSKI et al. 1991). COLE et al. (1985) originally identified MCP as a third class, in addition to CR1 and CR2, of electrophoretically distinct, C3 binding, membrane proteins of human peripheral blood leukocytes. Initially termed gp45–70, to reflect its electrophoretic mobility on SDS-PAGE, the common name was changed to MCP when its cofactor activity was recognized (SEYA et al. 1986). MCP was hypothesized and subsequently found to belong to a family of structurally, functionally, and genetically related proteins collectively termed the regulators of complement activation (RCA) (DECORDOBA et al. 1984, 1985; HOLERS et al. 1985; HOURCADE et al. 1989). Recently, MCP was given its leukocyte differentiation number (CD46) and a monoclonal antibody (E4.3) was designated as the standard typing reagent (HADAM 1990; PURCELL et al. 1989a). This chapter will review our current

---

Howard Hughes Medical Institute Laboratories, Department of Medicine, Division of Rheumatology, Washington University School of Medicine, Box 8045, 4566 Scott, St. Louis, MO 63110, USA

understanding of the structure and function of MCP, emphasizing recent contributions made possible by molecular analyses (POST et al. 1991).

2 Function

The wide tissue distribution, affinity for C3b and C4b, and factor I-dependent cofactor activity strongly suggested in early investigations that MCP would be a regulatory protein of the complement system (HOLERS et al. 1985; SEYA et al. 1986; ATKINSON and FARRIES 1987). Further, MCP-bearing cells (such as T lymphocytes, platelets, epithelial cells, and fibroblast cell lines) did not rosette with C3b/C4b-coated cells or efficiently bind fluid phase C3b (ATKINSON and FARRIES 1987; HOLERS et al. 1985; SEYA and ATKINSON 1989). This indicated that MCP was not a complement receptor. Coupling the latter findings with the similarity in cell and tissue distribution of MCP and another inhibitory protein of the complement cascade, decay accelerating factor (DAF) (see Chap. 2), led to the speculation that MCP would be an intrinsically acting regulator, i.e., it would primarily regulate C3b and C4b deposited on the same cell to which the MCP molecule was anchored (SEYA and ATKINSON 1989). Moreover, MCP had a complementary regulatory activity to DAF, i.e., DAF possessed decay accelerating activity but no cofactor activity, while MCP's activity profile was the opposite (LUBLIN and ATKINSON 1989b). It was anticipated, therefore, that DAF and MCP would act jointly to inhibit C3b/C4b deposition on self-tissue.

The cofactor activity of MCP for C3b and C4b was initially observed with solubilized preparations of cells bearing MCP as the only known cofactor protein (TURNER 1984; SEYA et al. 1986) and later with the purified MCP protein (SEYA et al. 1986). The demonstration of this same activity profile on the membrane of cells awaited the development of antibodies which blocked its function and transfection experiments with MCP cDNA. With these tools in hand, MCP has now been shown, in two separate demonstrations, to protect cells from complement-mediated attack. In one experimental system (SEYA et al. 1990b), a monoclonal antibody (mAb) to MCP that abrogates its ligand binding and cofactor activity was incubated with DAF-negative T cell lines. These lines subsequently became coated with C3b via the alternative pathway. Also, if the T cell line was also deficient in CD59 (an inhibitor of the membrane attack complex; see Chap. 8), the cells were also lysed. In the other experimental system, Chinese hamster ovary (CHO) and NIH 3T3 cells were transfected with MCP cDNA. These MCP-expressing cells were protected from lysis by the alternative or the classical complement pathway (ATKINSON et al. 1991). A mAb to MCP that blocks its ligand binding reversed its protective effect. These data directly demonstrate that MCP protects cells from C4b/C3b deposition.

The physiologic activation or "tickover" of C3b (and possibly also of C4b) is a continuous but low-grade process (LACHMANN and NICOL 1973). Such "activated" C3b and C4b deposit indiscriminately on targets (self as well as nonself). Much of the "activated" C3b and C4b never bind to targets but are inactivated by condensation with H_2O in the fluid phase (C3b(H_2O)). However, C3b and C4b on targets or in plasma are still hemolytically active since they can form convertases. The goal of the RCA system is to control the complement cascade at the critical level of C3 activation (HOLERS et al. 1985; ATKINSON and FARRIES 1987; HOURCADE et al. 1989). To accomplish this, decay accelerating and cofactor activities for C3b, C4b, and the two C3 convertases (C4bC2a and C3bBb) are necessary. In plasma (fluid phase), this regulatory problem has been solved by the hepatic synthesis of two proteins, factor H and C4 binding protein (C4bp), each of which binds one of the two possible substrates with high efficiency. However, factor H and C4bp possess both decay accelerating and cofactor activity for their respective ligands and for the convertase containing the specific ligand. In contrast, the division of labor is different on cells. DAF and MCP each bind both ligands but only possess one of the two regulatory activities. This symmetry relative to the overall activity profile of membrane vs fluid phase regulatory proteins of the C3 convertases is noteworthy (Table 1).

Another interesting feature of MCP is that it appears to be most effective against C3b covalently bound to large membrane proteins, including the C5 convertase (SEYA et al. 1991). The C5 convertase of the alternative pathway may consist of dimeric C3b (i.e., C3bBbC3b) (KINOSHITA et al. 1988). SEYA et al. (1991) found that degradation of only one of the two C3bs to C3bi by MCP and factor I was sufficient to inactivate the convertase. These data demonstrate an interesting parallel between the function of DAF and MCP. DAF prevents the assembly of convertases by binding to C4b or C3b but has an even higher affinity for the convertase as it begins to form (C4bC2a or C3bBb) (FUJITA et al. 1987). MCP can bind monomeric C4b and C3b but acts more efficiently on a C3b dimer, including C3bC3b in the form of C5 convertase (SEYA et al. 1991), and, presumably, for a C4b–C3b dimer as well.

Table 1. A comparison of the regulatory profile of the plasma and membrane inhibitors of C4 and C3

	C4b[a]		C3b[a]	
	Decay accelerating activity	Cofactor activity	Decay accelerating activity	Cofactor activity
Plasma				
C4bp	+	+	−	−
Factor H	−	−	+	+
Membrane				
DAF	+	−	+	−
MCP	−	+	−	+

[a] This activity profile applies to C4b and C3b as isolated proteins or as part of the classical or alternative pathway C3 and C5 convertases

Several additional points relative to the function of MCP are of interest. First, MCP, in the absence of factor I, stabilizes the alternative and classical pathway C3 convertases (SEYA and ATKINSON 1989). Although this observation establishes that MCP can bind C3b or C4b in the convertase, the physiologic significance of this finding is unclear since, in vivo, factor I would cleave the C3b or C4b. Second, fluid phase MCP has a low affinity for C3b on a cell, and cell bound MCP has limited ability to interact with C3b in the fluid phase (SEYA and ATKINSON 1989). These results are expected for an intrinsically acting regulatory protein. While these data parallel that of DAF (reviewed in LUBLIN and ATKINSON 1989a, 1991), they are in contrast to those observed with CR1 and factor H. Third, MCP is estimated to be 50 times more potent than factor H as a cofactor if both MCP and C3b are in the fluid phase (SEYA et al. 1986). Perhaps these experiments (with NP-40 present) mimic more closely the in vivo condition, in which C3b is bound to the same cell as MCP is anchored. Fourth, a formal demonstration that MCP acts exclusively as an intrinsic cofactor remains to be accomplished; such an experiment would need to show that MCP has cofactor activity primarily for C3b bound to the same cell on which the MCP is anchored and not for C3b bound to an adjacent cell.

Another area to be explored is the manner in which DAF and MCP may interact on a host cell being attacked by complement. For example, if C3b or C4b is deposited on a cell bearing DAF and MCP, does DAF arrive first and then MCP? Can DAF and MCP bind to the same C3b or C4b or do they compete for binding? Are they synergistic in their ability to inhibit C3b deposition on a target? Is there a critical number of MCP or DAF molecules on a membrane that are required, or can protection be enhanced in some proportion to the number of molecules per cell?

The functional profile of MCP is also of interest relative to autoimmunity, tumor immunity, and transplantation immunity. Under normal circumstances, only trace quantities of C3b/C4b are deposited on host tissue and these are promptly inactivated by DAF and MCP. Moreover, in most inflammatory conditions, even in the presence of ongoing complement activation, as during many types of infectious illnesses, damage to self-tissue is minimal. This self-protective system can be overcome, however. A common problem in autoimmunity is the synthesis of autoantibodies which activate the complement system and thereby have the potential to inflict damage to autologous cells and tissues. For example, in the cold agglutinin syndrome, an IgM autoantibody is made to an erythrocyte membrane molecule. In this case, there is enough complement activation to override the inhibitory activities of the regulatory proteins and the cells are lysed or phagocytosed. Conceivably, increasing the expression of DAF and/or MCP could be a feasible therapy in such autoimmune syndromes. Unfortunately, we know little about the regulation of expression of MCP or DAF. The quantification of these two proteins in autoimmune and other chronic inflammatory conditions would be a useful undertaking.

In tumor immunity, an attractive therapeutic approach would be to destroy tumor cells with antibody and host complement. However, the expression of

MCP appears to be increased on many malignant cells (SEYA et al. 1990a; CHO et al. 1991). Consequently, for C3b deposition to occur on tumor cells, this "protective" system must be overcome. If one could selectively reduce the expression or block the activity, as with an anti-MCP mAb, it should be possible to increase a tumor cell's susceptibility to complement-mediated destruction.

These regulatory proteins also account for the manner in which the complement system had handled the tricky question of separating self from non-self (ATKINSON and FARRIES 1987). The alternative pathway is an independent immune system. It is not only ancient on the phylogenetic scale but also almost certainly preceded the appearance of antibody (FARRIES et al. 1990; FARRIES and ATKINSON 1991). The alternative pathway can amplify on foreign but not host tissue. Microbes, for the most part, do not possess complement regulatory proteins. As a result, if C3b or C4b is deposited on a microbe, massive and rapid complement activation occurs on the target. No such activation occurs on self-tissue. This self/non-self discrimination in the complement system is made possible by the regulatory proteins. It is a remarkably simple, yet highly effective solution to a difficult problem.

The fact that the vaccinia virus (VV) and the herpes simplex viruses (HSV) express C4b and C3b binding proteins provides further evidence for the importance of regulatory proteins. By mimicking human proteins, these viral proteins serve as virulence factors in preventing complement activation. They are structurally and functionally related to the RCA group. The VV protein is most homologous to C4bp (KOTWOL and MOSS 1988). It is a secreted protein which binds to C4b and inhibits the classical pathway (KOTWAL et al. 1990). In the case of HSV, a glycoprotein, heavily expressed on the viral envelope and on virally infected cells (FRIEDMAN et al. 1984), acts as an inhibitor of the alternative pathway C3 convertase (FRIES et al. 1986). It renders infected cells more resistant to complement-mediated attack (MCNEARNEY et al. 1987).

Complement regulatory proteins also may have a role in transplantation immunity. Currently, xenografts are poorly tolerated and usually are rapidly rejected. Human organs do not meet the need. For a variety of reasons, a primate source is unlikely to become an acceptable alternative. Normally, xenografts are acutely rejected because of heterophile antibodies activating the classical complement pathway on the endothelium of a transplanted organ's vasculature (PLATT et al. 1990). In some cases the alternative complement pathway itself may mediate the rejection (ATKINSON et al. 1991). By expressing regulatory proteins such as MCP and/or DAF on graft endothelium, in theory it should be possible to prevent the acute graft rejection. Recent evidence suggests that pig endothelial cells coated with DAF (PLATT et al. 1990) or 3T3 cell lines expressing MCP or DAF are protected from this type of injury (ATKINSON et al. 1991). Transgenic animals expressing human DAF and/or MCP would provide the experimental system to more rigorously examine these ideas.

Another approach is to infuse such regulatory proteins to prevent complement-dependent tissue injury. Recently, a soluble form of human CR1 that possesses decay accelerating and cofactor activities was given intravenously before inducing a heart attack in a rat (WEISMAN et al. 1990). Ischemic reperfusion cardiac injury was inhibited. Also, injection of CR1 at the site of an Arthus reaction reduced the vasculitic insult (YEH et al. 1991). These initial studies suggest that complement regulatory proteins may become useful therapeutic agents.

3 Structure

3.1 Phenotypic Patterns

A distinguishing structural characteristic of MCP on SDS-PAGE is the presence of two heterogeneous protein species with M_rs of 59 000–68 000 and 50 000–58 000 (COLE et al. 1985). BALLARD et al. (1987) noted three patterns of expression of these two forms on human peripheral blood cells: (1) upper form predominant, 65% of the population; (2) approximately equal distribution of each band, 29%; and (3) lower form predominant, 6%. The protein phenotype was stable over time and the pattern was identical on an individual's B cells, T cells, monocytes, and platelets (YU et al. 1986; BALLARD et al. 1987; SEYA et al. 1988). On granulocytes (SEYA et al. 1988) MCP was expressed as a single broad band, possibly secondary to a different pattern of glycosylation that caused the two forms to overlap. The typical phenotypic patterns were observed, however, on tumor cell lines of hematopoietic, epithelial, and fibroblast lineages (MCNEARNEY et al. 1989). This finding indicates that a single cell expresses both forms of the protein. These data and family studies provide considerable evidence for an autosomal codominant inheritance pattern regulating the expression of the two forms of MCP (BALLARD et al. 1987). BORA et al. (1991) found that a *Hind*III restriction fragment length polymorphism (RFLP) correlated with this pattern of expression, indicating that a *cis*-acting factor is responsible. This particular polymorphic site was traced to an intron between exons 1 and 2, nearly 10 kb from the alternatively spliced exon 8 (see below). It is likely that this polymorphic *Hind*III site is associated with another change in the DNA in the introns surrounding exon 8.

3.2 Biosynthesis

Prior to molecular studies, MCP was characterized biochemically in order to provide an explanation of the two forms. In independent studies, BALLARD et al. (1988) and STERN et al. (1986) found that the upper and lower forms possessed N- and O-linked sugars. The shift in molecular weight following deglycosylation was most consistent with the presence of three N-linked complex carbohydrate units on the mature forms of the molecule and multiple (5–10) O-linked units.

Endoglycosidase digestions also demonstrated that the upper band protein contained more sialic acid. Further, much of the heterogeneity within the two forms was secondary to the variability in glycosylation (BALLARD et al. 1988). Also, consistent with the latter interpretation was the fact that the precursors (pro-MCP) were not broad species like the mature forms.

In the U937 monocyte cell line, two precursors with M_rs of 41 000 and 43 000 were identified (BALLARD et al. 1988). These precursors possessed N-linked high mannose carbohydrates and thus were processed in a pre-Golgi compartment. One precursor chased into the mature form with a $t_{1/2}$ of 20 min, while the other required 90 min. It was not possible to demonstrate that the lower molecular weight precursor chased into the lower molecular weight mature species. In fact, cell lines such as U937 and K562, with a predominant upper band phenotypic pattern of MCP expression, demonstrated two precursors of about equal intensity. Molecular analyses have now provided an explanation for these results (see below).

4 Sequence Analysis and Genomic Organization

4.1 cDNA Isoforms

SEYA et al. (1986) purified MCP and obtained an NH_2-terminal sequence. An oligonucleotide probe based on this sequence was used to clone an MCP cDNA (LUBLIN et al. 1988). The latter encoded for a 384 amino acid protein, including a 34 amino acid signal peptide. Beginning at the NH_2-terminal, MCP was composed of 4 of the approximately 60 amino acid cysteine-rich repeating modules known as short consensus repeats (SCRs); an area enriched in serines, threonines, and prolines (STP); a short segment (12 amino acids) of unknown significance (UK); a hydrophobic membrane-spanning domain with a basic cytoplasmic anchor (HY); and a cytoplasmic tail (CYT). Thus, MCP was a typical type 1 membrane glycoprotein (Fig. 1).

POST et al. subsequently characterized additional MCP cDNA clones (1991). These were identical except for two areas of variability, the STP-enriched domain and the CYT. The STP domain consisted of three possible segments, A, B, and C. Clones were found with one of four patterns relative to this area: STP^{ABC}, STP^{BC}, STP^B, or STP^C. In addition, a CYT of 16 amino acids (CYT^1) was discovered which differed from the originally identified tail of 23 amino acids (CYT^2). Each tail existed with any of the four varieties of STP regions.

Consequently, MCP was found to occur as a family of isoforms (Fig. 2). Polymerase chain reaction (PCR) and northern blot analyses revealed that cell lines and peripheral blood cells regularly expressed variable quantities of four of these isoforms (POST et al. 1991). Transfection studies determined that the higher molecular weight (upper band) pattern phenotype corresponded to isoforms

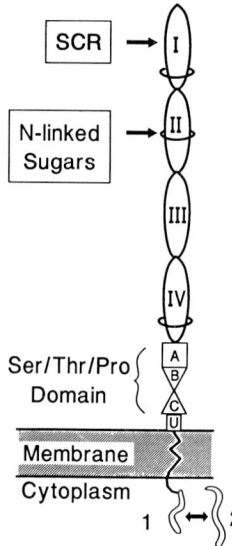

Fig. 1. Structural highlights of membrane cofactor protein. Its extracellular amino portion begins with four short consensus repeats (*SCR*) that function in ligand binding and cofactor activity. There are three potential sites for N-linked glycosylation. The area near the membrane (*Ser/Thr/Pro domain*) is a site of heavy O-linked glycosylation. Variability in the content of this area produces phenotypic changes (see text). Prior to the membrane insertion site is an area of unknown functional significance (*U*). At the COOH-terminal MCP possesses two distinct cytoplasmic tails that may be involved in differential intracytoplasmic processing of the protein

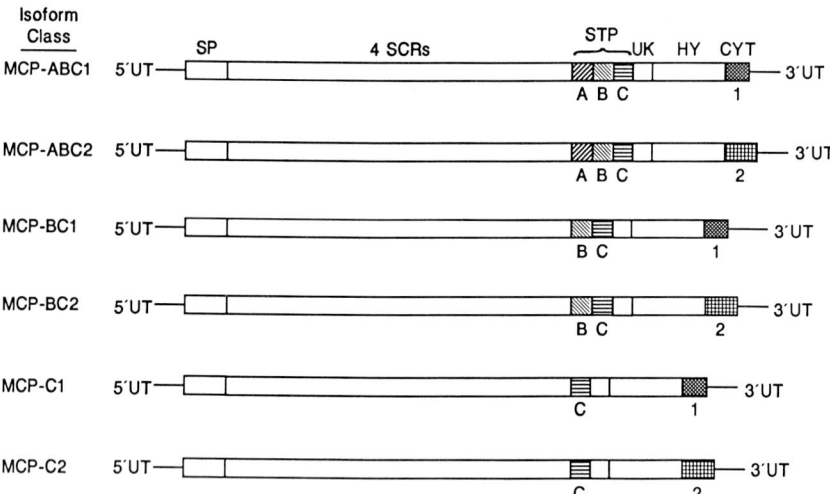

Fig. 2. Multiple MCP cDNAs arise by alternative splicing at the COOH-terminal. Six MCP cDNA classes arise from alternative splicing of the STP-enriched segments and cytoplasmic tails. Abbreviations described in Fig. 3

containing $STP^{BC} + CYT^1$ and $STP^{BC} + CYT^2$ (POST et al. 1991). The lower molecular weight (lower band) pattern correlated with isoforms bearing $STP^C + CYT^1$ and $STP^C + CYT^2$. Isoforms with STP^A were rare species in the cells and cell lines examined.

As a result of these studies (POST et al. 1991), it was established that the presence of the region termed STP^B correlated with the upper band phenotype and its absence with the lower band phenotype. Although the STP^B segment is only 15 amino acids in length, the addition of sugars to the OH-groups on serines and threonines would account for the molecular weight difference. Subsequently, POST et al. (1991) determined the genomic organization of the MCP gene and demonstrated that the isoforms arose by alternative splicing.

4.2 Genomic Organization

The MCP gene is situated on the long arm of chromosome 1 at q3.2 within the RCA locus (LUBLIN et al. 1988). MCP and four other RCA members lie on an 800 kb fragment in the order: MCP–CR1–CR2–DAF–C4bp (BORA et al. 1989). Factor H is located within the RCA cluster as well, but has not yet been linked at the DNA level. Of interest, Factor XIIIB of the clotting system has ten SCR domains and is tightly linked to the factor H gene (REY-CAMPOS et al. 1990).

The MCP gene consists of 14 exons and 13 introns and has a minimum length of 43 kb (see Fig. 3 and POST et al. 1991). Features of interest include:

1. A "split" exon encoding SCR-2. This occurs at the second nucleotide of glycine 34. At least one such split exon is present at the same nucleotide position in the other four RCA proteins.
2. Three closely spaced exons encoding the STP region. These exons immediately follow the exon for SCR-4. A comparison of the intronic and exonic sequences of STP^A and STP^B indicate that they arose by a duplication event. Presumably, an alteration in the introns surrounding STP^A regulates the efficiency of its splicing and thereby accounts for the expression polymorphism of MCP (BORA et al. 1991).

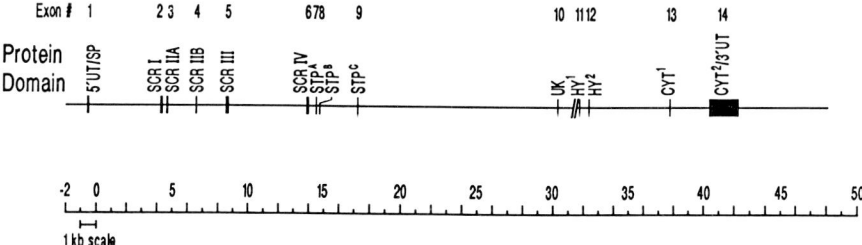

Fig. 3. Structure of the human MCP gene. The exon number and corresponding domain are shown above each exon. Intron and exon sizes are drawn to scale. As indicated by the *slashes* the intron between exons 10 and 11 has has not been cloned. The *lower line* indicates the scale. *5'UT/SP*, 5' untranslated area/signal peptide; *SCR*, short consensus repeat; *STP^A*, *STP^B*, *STP^C* serine/threonine/proline-rich areas; *UK*, region of unknown functional significance; *HY^1*, hydrophobic region one; *HY^2*, hydrophobic region two and amino acid anchor; *CYT^1*, cytoplasmic tail one; *CYT^2/3'UT*, cytoplasmic tail two and 3' untranslated area

3. A large gap of ~13 kb from the last STP exon to the exons encoding the UK and HY domains.
4. Two CYTs are alternatively spliced. Exon 13 encodes for CYT^1 and, when present, converts CYT^2 (within exon 14) into the 3'UT.
5. Alternative splicing of exons STP^B and CYT^1 gives rise to the four isoforms that are regularly expressed on human cells.

5 Correlation of Structure and Function

The SCRs are the ligand binding region of RCA proteins. Two SCRs form a binding site for C3b or C4b in CR1 (KLICKSTEIN et al. 1988; KRYCH et al. 1991) and for C3dg in CR2 (LOWELL et al. 1989; CAREL et al. 1990). For MCP, therefore, C3b and C4b binding sites would be expected within the SCRs. ADAMS et al. (1991), by deleting one SCR and then expressing the mutant protein in COS cells, have recently show that SCR-3 and SCR-4 are critical for C3b binding while SCR-2, -3, and -4 are necessary for C4b binding. These and other data suggest that the C3b and C4b binding sites are distinct and that SCR-2 contributes more to C4b than to C3b binding. Of much interest, a mutant lacking SCR-2 bound C3b but possessed no cofactor activity. These results not only suggest that ligand binding and cofactor activity can be separated but also that there is a direct interaction between a region within SCR-2 and factor I.

The role of the STP-rich region is unknown. REDDY et al. (1989) have suggested that similar regions in other membrane proteins may serve to protect the molecule from proteases. Others have suggested that these regions may serve as a means to induce the peptide core to adopt a linear and rigid confirmation (JENTOFT 1990). Due to the variability in its STP regions, MCP may provide a model system to examine the biological significance of this structural feature.

Interestingly, MCP on sperm appears to have no N- or O-linked carbohydrate units (CERVONI et al. 1991). It is not clear if this represents an alternatively spliced form lacking STP exons or a deficiency of the enzymes required for formation and attachment of sugars in this specialized tissue. However, the latter is suggested by the lack of both N- and O-linked units. Despite this carbohydrate deficit, sperm MCP possesses cofactor activity although its affinity for C3b is reduced compared to MCP of mononuclear cells.

The reason for two distinct CYTs is also unknown. Most cells express MCP having both types of CYTs, so this variation is likely to be of functional importance. MCP derived from solubilized transfected cells expressing one or the other CYTs showed no differences relative to cofactor activity (LISZEWSKI and ATKINSON, unpublished). Since ligand binding and cofactor activity are mediated solely by SCRs, the variability in STP-rich regions and CYTs must relate to other biological properties of the protein. For example, the two CYTs influence the processing of the pre-Golgi MCP precursor. In several transfected cell lines, MCP bearing CYT^1 arrives at the plasma membrane three to five times faster than MCP bearing CYT^2 (LISZEWSKI and ATKINSON, unpublished). It is possible that CYT^1 possesses a positive signal motif in its domain which facilitates its movement to the Golgi apparatus or that CYT^2 possesses a motif that delays its movement from the endoplasmic reticulum to the Golgi.

6 Discovery and Distribution

C3 affinity chromatography of solubilized human leukocyte preparations led to the discovery of MCP (COLE et al. 1985). This ligand binding activity, and especially its cofactor activity, provided a means to follow column fractions and thereby develop a scheme to purify the protein (SEYA et al. 1986). The purified protein was utilized to obtain NH_2-terminal sequence and for preparation of antibodies (SEYA et al. 1986; LUBLIN et al. 1988). After LUBLIN et al. (1988) published the derived amino acid sequence, it became apparent, through the work of PURCELL et al. (1989a, 1990a) and CHO et al. (1991), that MCP had also been identified at the protein level by other investigators.

During the 1980s, four other research groups independently produced antibodies to an antigen that was subsequently shown to be MCP. JOHNSON and colleagues in Liverpool (1981) produced a mAb to human syncytiotrophoblast microvilli. The antibody, H316, precipitated two sialated glycoproteins of trophoblast tissue and human lymphocytes (STERN et al. 1986) and a single band of 50 000 kDA from human sperm (ANDERSON et al. 1989). Unfortunately, this mAb is in limited supply and the hydridoma cell line has apparently been lost. More recently, this group has identified RFLPs of the MCP gene that may occur less frequently in individuals with recurrent spontaneous abortions (RISK et al. 1991).

HSI'S group in Nice, France also produced a mAb, GB24, to trophoblast tissue that recognized an antigen of lymphocytes and many other cell lines (FENICHEL et al. 1989; HSI et al. 1991). In addition, GB24 detected this antigen on the inner acrosomal membrane of human sperm (FENICHEL et al. 1989).

In Australia, SPARROW and MCKENZIE (1983) developed a monoclonal antibody (E4.3) using human lymphocytes as the immunogen. They subsequently demonstrated that E4.3 reacted with a membrane antigen on almost all human cells and cell lines tested and that was present in several body fluids (SPARROW and MCKENZIE 1983; PURCELL et al. 1990a). This group subsequently purified the protein and demonstrated that the NH_2-terminal sequence was identical to that of MCP (PURCELL et al. 1990b). In addition, they previously pointed out possible associations of MCP with HLA molecules (SPARROW and MCKENZIE 1983) and with envelope proteins of primate retroviruses (PURCELL et al. 1989a).

At the Wistar Institute, ANDREWS and colleagues (1985) immunized mice with a melanoma cell line and produced a mAb that reacted with a membrane protein on most human cells. They further demonstrated that the gene coding for this protein was on chromosome 1.

Due to the physical similarities of the protein identified by the various mAbs and the tissue distribution of the antigen, the Australian, French, and English groups had previously exchanged reagents. PURCELL et al. (1990a, b) and CHO et al. (1991) subsequently have showed that these mAbs all reacted with MCP. This finding, in addition to the amino acid sequence identity noted by the Australian group, established that the antigen recognized in these independent investigations was MCP.

The data accummulated by five groups point out the very broad tissue distribution of MCP. Except for erythrocytes, it has been found on nearly every cell and tissue examined: cells of fibroblast, epithelial, and endothelial lineages, including malignant cell lines such as HeLa, HEp-2, and transformed fibroblasts (MCNEARNEY et al. 1989); peripheral blood leukocytes (COLE et al. 1985; SEYA et al. 1988; SPARROW et al. 1983); platelets (YU et al. 1986; SEYA et al. 1988); sperm (FENICHEL et al. 1989; ANDERSON et al. 1989); trophoblast (STERN et al. 1986; HUNT and HSI 1990; HSI et al. 1991); and in serum, saliva, and seminal fluid (PURCELL et al. 1990a).

Quantification of MCP has been done by SEYA et al. (1990a) and CHO et al. (1991), who found that peripheral blood leukocyte populations expressed 5 000–15 000 copies of MCP/cell while hematopoietic lines expressed 30 000–70 000 MCP/cell. Additionally, some epithelial cell lines expressed 100 000–500 000 copies/cell.

Of special interest is the presence of MCP on syncytial trophoblasts (STERN et al. 1986; HUNT and HSI 1990; HSI et al. 1991) and the inner acrosomal membrane of sperm (FENICHEL et al. 1989; ANDERSON et al. 1989). On placental tissue, MCP may function to protect this "graft" from attack by the humoral immune system (HUNT and HSI 1990). Since the inner acrosomal membrane of sperm is not exposed until after it has contacted the egg, the function of MCP in this location is enigmatic. However, it has been proposed that it may be a strategy used by reproductive tissue to evade the humoral immune system.

7 MCP of Other Species

Relatively little is known about the phylogeny of complement membrane regulatory proteins. Orangutan MCP has been identified by affinity chromatography utilizing either human or orangutan iC3 as the ligand (NICKELLS and ATKINSON 1990). This orangutan protein was also immunoprecipitated by the human MCP mAb E4.3. Like the human protein, on SDS-PAGE it appeared as two forms. In contrast to orangutan, MCP was not found on gorilla erythrocytes.

Based on the characteristic, two band, electrophoretic pattern and affinity for rabbit iC3, MANTHEI et al. (1988) tentatively identified MCP on rabbit platelets. A number of other C3b binding proteins with a molecular weight similar to MCP have been identified on mouse cells (WONG and FEARON 1985), on marmoset erythrocytes (GOUJET-ZALC et al. 1987) and on baboon erythrocytes (BIRMINGHAM and COSIO 1989). In the case of baboon erythrocytes, NH_2-terminal sequencing of the protein indicated that it is highly homologous to the NH_2-terminal of human CR1 (BIRMINGHAM and COSIO 1991). The other proteins, then, could represent MCP, a smaller form of CR1, or a novel C3b binding protein.

8 Summary

MCP serves to down-regulate the activation of complement on host tissue. It performs this function by serving as a cofactor for the factor I-mediated cleavage of C3b and C4b. MCP is most likely an intrinsic regulator, i.e., it primarily protects its home cell. The wide tissue distribution of MCP mirrors this critical function of host cell protection. With the exception of erythrocytes, every cell and tissue examined expresses this protein.

MCP is represented as two broad heterogeneous bands on SDS-PAGE with M_rs of 51 000–58 000 and 59 000–68 000. The quantity of each form expressed is inherited in an autosomal codominant fashion. In most cells and cell lines, four isoforms of MCP predominate and arise by alternative splicing of a single MCP gene. All forms possess four repeating modules of—60 aminoacids, an area enriched in serines, threonines, and prolines [(STP), probable site of O-linked glycosylation], a short area of unknown function, a transmembrane domain, and a cytoplasmic tail. The isoforms differ, however, in the length and composition of the STP region and in the cytoplasmic tail. Alternative splicing of a single exon within the STP region determines the protein phenotype. Alternative splicing at the COOH-terminus gives rise to two distinct cytoplasmic tails. The biological significance of these structural variations in the STP and cytoplasmic tail regions is being investigated.

References

Adams EM, Brown MC, Nunge M, Krych M, Atkinson JP (1991) Contribution of the repeating domains of membrane cofactor protein (MCP; CD46) of the complement system to ligand binding and cofactor activity. J Immunol 147: 3005–3011

Anderson DJ, Michaelson JS, Johnson PM (1989) Trophoblast/leukocyte-common antigen is expressed by human testicular germ cells and appears on the surface of acrosome-reacted sperm. Biol Reprod 41: 285–293

Andrews PW, Knowles BB, Parkar M, Pym B, Stanley K, Goodfellow PN (1985) A human cell-surface antigen defined by a monoclonal antibody and controlled by a gene on human chromosome 1. J Ann Hum Genet 49: 31–39

Atkinson JP, Farries T (1987) Separation of self from non-self in the complement system. Immunol Today 8: 212–215

Ballard L, Seya T, Teckman J, Lublin DM, Atkinson JP (1987) A polymorphism of the complement regulatory protein MCP (membrane cofactor protein of gp45–70). J Immunol 138: 3850–3855

Ballard LL, Bora NS, Yu GH, Atkinson JP (1988) Biochemical characterization of membrane cofactor protein of the complement system. J Immunol 141: 3923–3939

Birmingham DJ, Cosio FG (1989) Characterization of the baboon erythrocyte C3b-binding protein. J Immunol 142: 3140–3144

Birmingham DJ, Cosio FG, McCourt DW, Atkinson JP (1991) N-terminus amino acid analysis of the baboon erythrocyte C3b receptor. FASEB 5: JA1716

Bora NS, Lublin DM, Kumar BV, Hockett RD, Holers, VM, Atkinson JP (1989) Structural gene for human membrane cofactor protein (MCP) of complement maps to within 100 kb of the 3' end of the C3b/C4b receptor gene. J Exp Med 169: 597–602

Bora NS, Post TW, Atkinson JP (1991) Membrane cofactor protein (MCP) of the complement system: A Hind III RFLP that correlates with the expression polymorphism. J Immunol 146: 2821–2825

Campbell RD, Law SKA, Reid KBM, Sim RB (1988) Structure, organization and regulation of the complement genes. Annu Rev Immunol 6: 161–195

Carel JC, Myones BL Frazier B, Holers VM (1990) Structural requirements for C3d,g/Epstein-Barr virus receptor (CR2/CD21) ligand binding, internalization, and viral infection. J Biol Chem 265: 12293–12299

Cervoni F, Oglesby TJ, Nickells M, Milesi-Fluet C, Fenichel P, Atkinson JP, Hsi B-L (1992) Identification and characterization of membrane cofactor protein (MCP) of human spermatozoa. J Immunol (in press)

Cho S-W, Oglesby TJ, Hsi B-L, Adams EM, Atkinson JP (1991) Characterization of three monoclonal antibodies to membrane cofactor protein (MCP) of the complement system and quantitation of MCP by radioassay. Clin Exp Immunol 83: 257–261

Cole J, Housley GA, Dykman TR, MacDermott RP, Atkinson JP (1985) Identification of an additional class of C3-binding membrane proteins of human peripheral blood leukocytes and cell lines. Proc Natl Acad Sci USA 82: 859–863

DeCordoba R, Dykman TR, Ginsberg-Fellner F, Ercilla G, Aqua M, Atkinson JP, Rubinstein P (1984) Evidence for linkage between the loci coding for the binding protein for the fourth component of human complement (C4bp) and for the C3b/C4b receptor. Proc Natl Acad Sci USA 81: 7890–7892

Decordoba SR, Lublin DM, Rubinstein P, Atkinson JP (1985) Human genes for three complement components that regulate the activation of C3 are tightly linked. J Exp Med 161: 1189–1195

Farries TC, Atkinson JP (1991) Evolution of the complement system. Immunol Today 12: 295–300

Farries TC, Steuer Knutzen KL, Atkinson JP (1990) Evolutionary implications of a new bypass activation pathway of the complement system. Immunol Today 11: 78–80

Fenichel P, Hsi B-L, Farahifar D, Donzeau M, Barrier-Delpech D, Yeh CJG (1989) Evaluation of the human sperm acrosome reaction using a monoclonal antibody, GB24, and fluorescence-activated cell sorter. J Reprod Fertil 897: 699–706

Friedman HM, Cohen GH, Eisenberg RJ, Seidel CA, Cines DB (1984) Glycoprotein C of herpes simplex virus 1 acts as a receptor for C3b complement component on infected cells. Nature 309: 633–637

Fries LF, Friedman HM, Cohen GH, Eisenberg RJ, Hammer CH, Frank MM (1986) Glycoprotein C of herpes simplex virus 1 is an inhibitor of the complement cascade. J Immunol 137: 1636–1640

Fujita T, Inoue T, Ogawa K, Iida K, Tamura N (1987) The mechanism of action of decay-accelerating factor (DAF). DAF inhibits the assembly of C3 convertases by dissociating C2a and Bb. J Exp Med 166: 1221–1228

Goujet-Zalc C, Ripoche J, Guercy A, Mahouy G, Fontaine M (1987) Marmoset red blood cell receptor for membrane associated complement components is not related to human CR1: Partial characterization of the C3-binding proteins responsible for the spontaneous rosette formation between marmoset red blood cells and human leukocytes. Cell Immunol 109: 282–294

Hadam MR (1990) Cluster Report: CD46. In: Knapp W (ed) Leukocyte typing IV. University of Oxford Press, Oxford, p 649

Holers VM, Cole JL, Lublin DM, Seya T, Atkinson JP (1985) Human C3b- and C4b-regulatory proteins: a new multi-gene family. Immunol Today 6: 188–192

Hourcade D, Holers VM, Atkinson JP (1989) The regulators of complement activation (RCA) gene cluster. Adv Immunol 45: 381–416

Hsi B-L, Jing C, Yeh G, Fenichel P, Samson M, Grivaux C (1988) Monoclonal antibody GB24 recognizes a trophoblast-lymphocyte cross-reactive antigen. Am J Reprod Immunol Microbiol 18: 21–27

Hsi B-L, Hunt JS, Atkinson JP (1991) Differential expression of complement regulatory proteins on subpopulations of human trophoblast cells. J Reprod Immunol 19: 209–223

Hunt JS, Hsi B-L (1990) Evasive strategies of trophoblast cells: selective expression of membrane antigens. Am J Reprod Immunol Microbiol 23: 57–63

Jentoft N (1990) Why are proteins O'glycosylated? Trends Biochem Sci 15: 291–294

Johnson PM, Cheng HM, Molloy CM, Stern CMM, Slade MB (1981) Human trophoblast-specific surface antigens identified using monoclonal antibodies. Am J Reprod Immunol 1: 246–254

Kinoshita T, Takata Y, Kozono H, Takada J, Hong K, Inoue K (1988) C5 convertase of the alternative pathway: covalent linkage between two C3b molecules within the trimolecular complex enzyme. J Immunol 141: 3895–3901

Klickstein LB, Bartow TJ, Miletic V, Rabson LD, Smith JA, Fearon DT (1988) Identification of distinct C3b and C4b recognition sites in the human C3b/C4b receptor (CR1, CD35) by deletion mutagenesis. J Exp Med 168: 1699–1717

Kotwal GJ, Moss B (1988) Vaccinia virus encodes a secretory polypeptide structurally related to complement control proteins. Nature 335: 176–178

Kotwal GJ, Isaacs SN, McKenzie R, Frank MM, Moss B (1990) Inhibition of the complement cascade by the major secretory protein of vaccinia virus. Science 250: 827–830

Krych M, Hourcade D, Atkinson JP (1991) Sites within the complement C3b/C4b receptor important for the specificity of ligand binding. Proc Natl Acad Sci USA 88: 4353–4357

Lachmann PJ, Nicol PAE (1973) Reaction mechanism of the alternative pathway of complement fixation. Lancet i: 465–467

Liszewski MK, Post TW, Atkinson JP (1991) Membrane cofactor protein (MCP or CD46): newest member of the regulators of complement activation gene cluster. Annu Rev Immunol 9: 431–455

Lowell AC, Klickstein LB, Carter RH, Mitchell JA, Fearon DT, Ahearn JM (1989) Mapping of the Epstein-Barr virus and C3dg binding sites to a common domain in complement receptor type 2. J Exp Med 170: 1931–1946

Lublin D, Atkinson JP (1989a) Decay accelerating factor: molecular biology, chemistry, and function. Annu Rev Immunol 7: 35–58

Lublin DM, Atkinson JP (1989b) Decay-accelerating factor and membrane cofactor protein. In: Lambris JD (ed) The third component of complement. Chemistry and biology. Springer, Berlin Heidelberg New York, pp 123–145 (Current topics in microbiology and immunology, vol 153)

Lublin D, Atkinson JP (1992) Decay accelerating factor and membrane cofactor protein. In: Sim RB (ed) Biochemistry and molecular biology of complement. MTP Press, Lancaster (in press)

Lublin DM, Liszewski MK, Post TW, Arce MA, LeBeau MM, Rebentisch MB, Lemons RS, Seya T, Atkinson JP (1988) Molecular cloning and chromosomal localization of human membrane cofactor protein (MCP): evidence for inclusion in the multi-gene family of complement-regulatory proteins. J Exp Med 168: 181–194

Manthei U, Nickells M, Barnes SH, Ballard LL, Cui W, Atkinson JP (1988) Identification of a C3b/iC3 binding protein of rabbit platelets and leukocytes: a CR1-like candidate for the immune adherence receptor. J Immunol 140: 1228–1236

McNearney TA, Odell C, Holers VM, Spear PG, Atkinson JP (1987) Herpes simplex virus glycoproteins gC-1 and gC-2 bind to the third component of complement and provide protection against complement-mediated neutralization of viral infectivity. J Exp Med 166: 1525–1535

McNearney T, Ballard L, Seya T, Atkinson JP (1989) Membrane cofactor protein of complement is present on human fibroblast, epithelial and endothelial cells. J Clin Invest 84: 538–545

Nickells MW, Atkinson JP (1990) Characterization of CR1 and membrane cofactor protein-like proteins of two primates. J Immunol 144: 4262–4268

Platt JL, Oglesby TJ, Vercellotti GM, Dalmasso AP, Matas AJ, Bolman RM, Najarian JS, Bach FH (1990) Transplantation of discordant xenografts: a review of progress. Immunol Today 12: 450–456

Post TW, Liszewski MK, Adams EM, Tedja I, Miller EA, Atkinson JP (1991) Membrane cofactor protein of the complement system: alternative splicing of serine/threonine/proline-rich exons and cytoplasmic tails produces multiple isoforms which correlate with protein phenotype. J Exp Med 174: 93–102

Purcell DFJ, Clark GJ, Brown MA, McKenzie IFC, Sandrin MS, Deacon NJ (1989a) HuLy-m5, an antigen sharing epitopes with envelope gp70 molecules of primate retroviruses has a structural relationship with complement regulatory molecules. In: Knapp W (ed) Leukocyte typing IV. University of Oxford Press, Oxford, pp 653–655

Purcell DFJ, McKenzie IFC, Lublin DM, Johnson PM, Atkinson JP, Oglesby TJ, Deacon NJ (1990a) The human cell-surface glycoproteins HuLy-m5, membrane cofactor protein (MCP) of the complement system, and trophoblast-leukocyte common (TLX) antigen are CD46. Immunology 70: 155–161

Purcell DFJ, Deacon NJ, Andrew SM, McKenzie IFC (1990b) Human non-lineage antigen CD46 (HuLy-m5): purification and partial sequencing demonstrates structural homology with complement regulating glycoproteins. Immunogenetics 31: 21–28

Reddy P, Caras I, Krieger M (1989) Effects of O-linked glycosylation on the cell surface expression and stability of decay accelerating factor, a glycophospholipid-anchored membrane protein. J Biol Chem 264: 17329–17336

Rey-Campos J, Baeza-Sanz D, de Cordoba SR (1990) Physical linkage of the human genes coding for complement factor H and coagulation Factor XIIIB subunit. Genomics 7: 644–666

Risk JM, Flanagan BF, Johnson PM (1991) Polymorphism of the human CD46 gene in normal individuals and in recurrent spontaneous abortion. Hum Immunol 30: 162–167

Seya T, Atkinson JP (1989) Functional properties of membrane cofactor protein of complement. Biochem J 264: 581–588

Seya T, Turner J, Atkinson JP (1986) Purification and characterization of membrane cofactor protein (gp45–70) which is a cofactor for cleavage of C3b and C4b. J Exp Med 163: 837–855

Seya T, Ballard L, Bora N, McNearney T, Atkinson JP (1988) Distribution of membrane cofactor protein (MCP) of complement on human peripheral blood cells. Eur J Immunol 18: 1289–1294

Seya T, Hara T, Matsumoto MG, Akedo A (1990a) Quantitative analysis of membrane cofactor protein (MCP) of complement. J Immunol 145: 238–245

Seya T, Hara T, Matsumoto M, Sugita Y, Akedo H (1990b) Complement-mediated tumor cell damage induced by antibodies against membrane cofactor protein (MCP, CD46). J Exp Med 172: 1673–1677

Seya T, Okada M, Matsumoto M, Hong K, Kinoshita T, Atkinson JP (1991) Preferential inactivation of the C5 convertase of the alternative complement pathway by factor I and membrane cofactor protein (MCP). Mol Immunol 28: 1137–1147

Sparrow RL, McKenzie IFC (1983) HuLy-m5: a unique antigen physically associated with HLA molecules. Hum Immunol 7: 1–15

Stern PL, Beresford N, Thompson S, Johnson PM, Webb PD, Hole N (1986) Characterization of the human trophoblast-leukocyte antigenic molecules defined by a monoclonal antibody. J Immunol 137: 1604–1609

Turner JR (1984) Structural and functional studies of the C3b and C4b binding proteins of a human monocyte-like cell line (U937) Masters thesis. Washington University, St Louis

Weisman HF, Bartow T, Leppo MK, Marsh Jr HC, Carson GR, Concino MF, Boyle MP, Roux KH, Weisfeldt ML, Fearon DT (1990) Soluble human complement receptor type 1: in vivo inhibitor of complement suppressing post-ischemic myocardial inflammation and necrosis. Science 249: 146–151

Oglesby TJ, White D, Tedja I, Liszewski K, Wright L, Van den Bogarde J, Atkinson JP (1992) Protection of mammalian cells from complement-mediated lysis by transfection of human membrane cofactor protein (MCP) and decay accelerating factor (DAF). Trans Assoc Amer Phys (in press)

Wong WW, Fearon DT (1985) p65: a C3b-binding protein on murine cells that shares antigenic determinants with the human C3b receptor (CR1) and is distinct from murine C3b receptor. J Immunol 134: 4048–4056

Yeh CG, Marsh Jr HC, Carson GR, Berman L, Concino MF, Scesney SM, Kuestner RE, Skibbens R, Donahue KA, Ip SH (1991) Recombinant soluble human complement receptor type 1 inhibits inflammation in the reversed passive arthus reaction in rats. J Immunol 146: 250–256

Yu G, Holers VM, Seya T, Ballard L, Atkinson JP (1986) Identification of a third component of complement-binding glycoprotein of human platelets. J Clin Invest 78: 494–501

Membrane Inhibitor of Reactive Lysis

M. H. HOLGUIN and C. J. PARKER

1	Introduction	61
2	Indirect Evidence of the Existence of Membrane Regulators of the MAC	62
2.1	Homologous Restriction of Complement-Mediated Lysis	62
2.2	Reactive Lysis of PNH Erythrocytes	63
3	Isolation of the Homologous Restriction Factor/C8 Binding Protein	64
4	Isolation of MIRL	65
5	Relationship Between MIRL and the Erythrocyte Phenotypes of PNH	68
6	Isolated Deficiencies of DAF and MIRL	69
7	Relationship Between MIRL and HRF/C8bp	70
8	Structural Characteristics of MIRL	70
8.1	Primary Sequence	70
8.2	Sequence Homology	72
8.3	Glycosylation	72
8.4	Sensitivity to Chemical Agents	72
8.5	Alternative Forms of MIRL	73
8.6	MIRL is GPI-Anchored	74
9	Cellular Expression of MIRL	76
9.1	Human Tissue	76
9.2	MIRL Expression by Cell Lines	76
10	Analysis of the Functional Properties of MIRL	77
10.1	Inhibition of Complement-Mediated Lysis	77
10.2	MIRL and Nonerythroid Cells	78
10.3	Homologous Restriction of Complement Activation	79
10.4	Effects on Early Stages of Complement Activation	80
10.5	Effects on Cell-Mediated Lysis	81
10.6	MIRL and T Cell Activiation	81
11	Conclusions	82
	References	82

1 Introduction

The membrane inhibitor of reactive lysis (MIRL, CD59) is an 18 kDa glycosyl phosphatidylinositol-anchored membrane glycoprotein that inhibits complement-mediated lysis, at least in part by restricting the formation of the membrane

---

Hematology/Oncology Section (111C), Veterans Administration Medical Center, 500 Foothill Drive, Salt Lake City, UT 84148, USA

attack complex (MAC). MIRL is expressed by all peripheral blood hematopoietic cells, by endothelial cells, and by some cellular elements of the peripheral and central nervous systems. The erythrocytes of paroxysmal nocturnal hemoglobinuria (PNH) are deficient in MIRL, and the enhanced susceptibility of PNH erythrocytes to complement-mediated lysis is due primarily to this deficiency.

2 Indirect Evidence of the Existence of Membrane Regulators of the MAC

Between 1977 and 1984, two areas of investigation produced evidence indicating that the lytic activity of complement is greatly influenced by the biochemical properties of the surface upon which the MAC is assembled. One set of studies focused on elucidating the mechanism of homologous restriction of complement-mediated lysis, a term used to describe the observation that cells are more efficiently lysed by heterologous complement than by homologous complement. A second set of studies was aimed at establishing the basis of the difference in susceptibility among the three phenotypes of PNH erythrocytes to reactive lysis.

2.1 Homologous Restriction of Complement-Mediated Lysis

YAMAMOTO (1977) observed that, after human C5b-8 complexes were formed on either guinea pig, mouse, sheep, or goat erythrocytes, addition of guinea pig C9 caused greater lysis of sheep and goat erythrocytes than of guinea pig or mouse erythrocytes. He further showed that the decreased capacity of guinea pig C9 to induce lysis of guinea pig or mouse erythrocytes was not due to a failure to bind to the human C5b-8 complex.

The observations of YAMAMOTO were confirmed and expanded by HÄNSCH et al. (1981), who reported that lysis of human, guinea pig, rabbit, mouse, and rat erythrocytes was least when the C8 and C9 were from the same species. Additional experiments suggested that the source of the C9 was more critical than the source of the C8.

Further studies designed to elucidate the basis of homologous restriction of complement-mediated lysis were performed by HU and SHIN (1984). They reported that C9 was inserted into the cell membrane much more efficiently on sheep than on human erythrocytes. Together, the studies of YAMAMOTO, HÄNSCH et al., and HU and SHIN provided compelling indirect evidence of the existence of an erythrocyte membrane constituent (or constituents) that restricted the lytic activity of the MAC by interfering with the functional activity of C9. The putative membrane inhibitor appeared to be most effective when the cells and the C9 were of homologous origins.

2.2 Reactive Lysis of PNH Erythrocytes

The erythrocytes of PNH have been classified into three groups based on susceptibility to complement-mediated lysis. Type I cells are normal or nearly normal in susceptibility; type II cells are moderately sensitive, requiring 1/3–1/5 as much serum (as the complement source) to produce an equal degree of lysis compared to normal erythrocytes; and type III cells are markedly sensitive, requiring 1/15–1/25 as much serum for an equal degree of lysis compared to normal cells (ROSSE and PARKER 1985). PNH III cells are susceptible to reactive lysis whereas PNH I, PNH II, and normal erythrocytes are resistant to this process (PACKMAN et al. 1979; PARKER et al. 1985).

In reactive lysis systems, neither the classical nor the alternative pathway is activated directly on the cell surface. Rather, isolated C5b-9 complexes are assembled on the cell membrane by either of two methods. One procedure employs purified components of the MAC. In this system, erythrocytes are incubated with isolated C5b-6 complexes along with C7. Binding of C7 to C5b-6 induces a conformational change in C7 such that the amphipathic molecule undergoes a hydrophilic to hydrophobic transition. This process endows the nascent C5b-7 complex with the capacity to insert into the lipid bilayer of the cell membrane and to form a stable intermediate (EC5b-7). Next, isolated C8 and C9 are added to form the cytolytic MAC.

Using the system described above, ROSENFELD et al. (1980) compared the sensitivity of normal and PNH III erythrocytes to reactive lysis using both human and guinea pig complement reagents. PNH III erythrocytes were much more susceptible to hemolysis than normal cells when human components were used, but normal erythrocytes became as susceptible as PNH erythrocytes when guinea pig complement components were used. Additional experiments suggested that the variation in reactive lysis sensitivity was due primarily to differences in the species of C9 used. These studies implied that PNH III erythrocytes are deficient in a membrane constituent that regulates the functional activity of human C9. That normal erythrocytes hemolyzed to the same degree as PNH III erythrocytes when guinea pig C9 was used suggested that the membrane constituent that controls human C9 activity does not inhibit guinea pig C9 activity. This latter observation was consistent with the paradigm of homologous restriction of complement-mediated lysis.

The second method used to induce reactive lysis takes advantage of the special properties of cobra venom factor (CoF). CoF is a structural analogue of human C3c (VOGEL et al. 1984), but it is functionally analogous to human C3b in that it can bind factor B. Unlike C3b, however, CoF is not subject to inactivation and degradation by factors H and I (LACHMANN and HALBWACHS 1975). The CoFB complex can be activated by factor D to form CoFBb. Depending upon the species from which it is derived, the CoF can bind either C3 or both C3 and C5. Consequently, the CoFBb complex will act as either a C3 convertase or both a C3 and a C5 convertase. The CoFBb complex is extremely stable (it has a half-life of approximately 7 h at 37°C) (VOGEL and MÜLLER-EBERHARD 1982), and, inasmuch as the CoF molecule is not susceptible to inactivation by the endogenous regulatory proteins, the complex is a potent activator of the alternative pathway of complement in serum. To insure that complement activation is mediated only by the CoFBb complex, serum is treated with EDTA, since the classical and alternative pathways are divalent cation-dependent. When incubated with CoFBb and serum containing EDTA (EDTA-serum), PNH III erythrocytes are hemolyzed but normal, PNH I, and PNH II erythrocytes are not. Quantitative analysis of the MAC components shows that PNH III cells bind much greater amounts of C5b-9 than normal cells (PARKER et al. 1985, 1989).

Detailed studies using the CoF system suggest that there are two steps at which PNH III erythrocytes fail to control MAC generation (PARKER et al. 1985, 1989). First, PNH III cells bind more C5b-7 than normal erythrocytes. This observation suggests that PNH III erythrocytes are deficient in a membrane constituent that regulates either the activity of the C3/C5 convertase or the assembly of the MAC at the stage of cell surface formation of the trimolecular C5b-7 complex. Second, quantitation of C9 indicates that PNH III cells lack a component that normally inhibits binding of multiple molecules of C9 to the C5b-8 complex. This latter observation can be interpreted as being consistent with the findings obtained using isolated components of the MAC to induce reactive lysis. As discussed above, those studies indicated that PNH III erythrocytes lack a constituent that controls the functional activity of C9.

Either of two hypotheses seemed plausible to explain the observations made using the CoF system. First, there could be two MAC regulatory constituents that are deficient in PNH. One would control either the generation of C5b (through regulation of the CoFBb C5 convertase) or the assembly of the C5b-7 complex, while the other would control the interactions of C9 with the C5b-8 complex. Second, PNH III erythrocytes could be deficient in a single constituent that controls both processes. Regardless of which hypothesis was correct, in order to be completely consistent with the experimental data, it was necessary to account for both the greater binding of C5b-7 and the aberrant interactions of C9 on PNH III erythrocytes.

3 Isolation of the Homologous Restriction Factor/C8 Binding Protein

In 1986, two groups working independently reported the isolation of a protein derived from human erythrocyte membranes that inhibited reactive lysis. The 65 kDa protein, named C8 binding protein (C8bp) by SCHÖNERMARK et al. (1986) and homologous restriction factor (HRF) by ZALMAN et al. (1986), appeared to inhibit the activity of the MAC by binding to C8 or C9 or both (a detailed description of the isolation and characterization of HRF/C8bp is contained in this volume). Subsequent studies also demonstrated that PNH III erythrocytes were deficient in HRF/C8bp (HÄNSCH et al. 1987; ZALMAN et al. 1987). Together, these results suggested that the enhanced susceptibility of PNH erythrocytes to reactive lysis was a consequence of HRF/C8bp deficiency. As discussed above, however, using the CoF system, PNH III erythrocytes had been shown to bind supranormal amounts of both C5b-7 and C9 (PARKER et al. 1985, 1989). Inasmuch as ZALMAN et al. (1986) had reported that HRF/C8bp had no effect on assembly of the C5b-7 complex, we postulated the existence of another erythrocyte membrane protein that inhibited complement-mediated lysis by regulating the activity of the MAC.

4 Isolation of MIRL

If our hypothesis was correct, normal erythrocytes would have a membrane protein that inhibited CoF-initiated hemolysis, but PNH erythrocytes would lack this inhibitor. To test the hypothesis, membrane proteins were extracted from normal and PNH erythrocytes using butanol (HOLGUIN et al. 1989a). Next, PNH III erythrocytes were incubated with incremental concentrations of the extract derived from either normal or PNH cells. After washing, the treated erythrocytes were incubated with activated CoF complexes (CoFBb) and a dilution of normal human serum sufficient to mediate partial lysis of the PNH III cells. The extract derived from normal erythrocytes inhibited CoF-initiated lysis of PNH eryth-

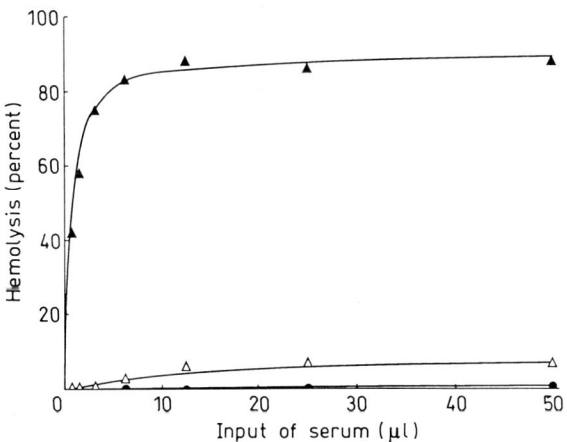

Fig. 1. MIRL inhibits reactive lysis of PNH III erythrocytes. Normal erythrocytes were incubated with buffer (●) and PNH III erythrocytes were incubated with buffer (△) or with buffer containing MIRL (5 µg/ml) (▲). After washing, the cells were incubated with cobra venom factor and incremental concentrations of human serum. Pretreatment with MIRL protected PNH III erythrocytes against cobra venom factor initiated lysis. (Reproduced from the *Journal of Clinical Investigation*, 1989, 84:12, by copyright permission of the American Society of Clinical Investigation)

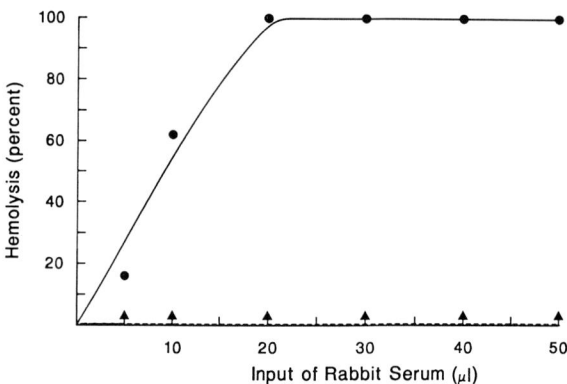

Fig. 2. Normal erythrocytes are rendered susceptible to cobra venom factor initiated hemolysis after treatment with anti-MIRL. Normal erythrocytes were incubated with incremental concentrations of anti-MIRL antiserum (●) or nonimmune rabbit serum (▲). After washing, the cells were incubated with cobra venom factor complexes and serum containing EDTA, and hemolysis was quantified. The data points represent the mean of triplicate determinations. Inhibition of MIRL function induces susceptibility to cobra venom factor initiated hemolysis. (Reproduced from the *Journal of Clinical Investigation*, 1989, 84:12, by copyright permission of the American Society of Clinical Investigation)

rocytes in a dose-dependent fashion, while the PNH extract had essentially no inhibitory activity (HOLGUIN et al. 1989a).

The inhibitor was isolated to apparent homogeneity by subjecting the butanol extract derived from normal erythrocyte membranes to sequential anion exchange, hydroxylapatite, and hydrophobic interactive chromatography (HOLGUIN et al. 1989a). Activity was followed using an assay that quantitated the capacity of the column fractions to inhibit CoF-initiated hemolysis of guinea pig erythrocytes. Analysis by SDS-PAGE and silver staining of a pool of the active fractions showed a single band representing a protein with an M_r of 18 kDa. Based on its capacity to inhibit CoF-initiated hemolysis of PNH erythrocytes (Fig. 1), the protein was called membrane inhibitor of reactive lysis (MIRL). Additional evidence that MIRL regulates susceptibility to reactive lysis was provided by the results of experiments in which the activity of MIRL on normal erythrocytes was blocked by using a monospecific antiserum. Anti-MIRL treatment caused a dose-dependent susceptibility to CoF-initiated lysis, while cells incubated with nonimmune rabbit serum were resistant to lysis (Fig. 2).

Further experiments showed that isolated MIRL remained functionally active after spontaneously incorporating into cell membranes (HOLGUIN et al. 1989a). Since this property is characteristic of glycosyl phosphatidylinositol (GPI)-linked proteins (MEDOF et al. 1984), the results provided indirect evidence that, like other proteins that are missing from PNH erythrocytes, MIRL belongs to the family of GPI-linked proteins. Analysis by western blot confirmed that PNH III erythrocytes lacked MIRL (Fig. 3; HOLGUIN et al. 1989a).

Working independently, three other groups isolated a protein identical to MIRL and demonstrated its complement regulatory activity (SUGITA et al. 1988;

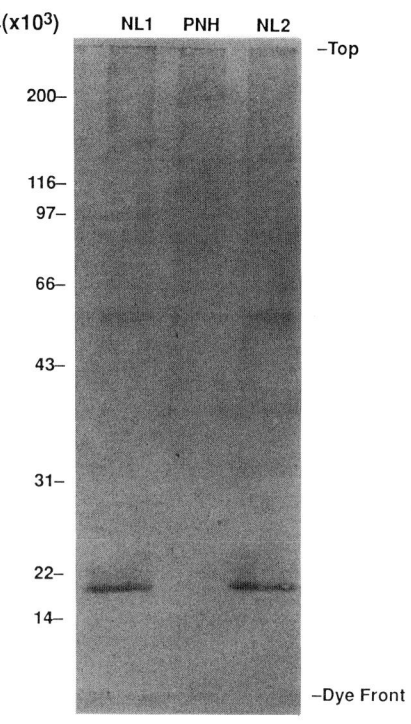

Fig. 3. Analysis of western blot of normal and PNH erythrocyte membrane proteins using anti-MIRL as the primary antibody. Five µg of hemoglobin-free erythrocyte ghosts were subjected to SDS-PAGE under nonreducing conditions and electrophoretically transferred to nitrocellulose paper. After incubation with anti-MIRL antiserum, antibody binding was localized by using alkaline phosphatase-conjugated anti-rabbit IgG and a chromogenic substrate. The PNH erythrocytes are deficient in MIRL. (Reproduced from the *Journal of Clinical Investigation*, 1989, 82:12, by copyright permission of the American Society of Clinical Investigation)

DAVIES et al. 1989; OKADA et al. 1989a). Two other groups have developed monoclonal antibodies against the MIRL protein, but they did not initially appreciate that the antigen recognized by their antibody functioned as a regulator of complement (STEFANOVA et al. 1989; GROUX et al. 1989).

Using conditions that were designed for the purification of HRF/C8bp, SUGITA et al. (1988) isolated an 18 kDa protein that inhibited reactive lysis of guinea pig erythrocytes, but an inhibitor with the reported properties of HRF/C8bp was not observed. In subsequent studies, SUGITA and colleagues have referred to the 18 kDa protein as membrane attack complex-inhibitory factor (MACIF) and have reported that it inhibits C9 binding, C9 polymerization, or both (SUGITA et al. 1989b).

OKADA and colleagues (1989a, c) developed a monoclonal antibody (1F5) that bound to neuraminidase-treated human erythrocytes and rendered them susceptible to complement-mediated lysis initiated by activation of the alternative pathway. By using affinity chromatography, the antigen recognized by 1F5 was isolated and estimated to have an Mr of 20 kDa (HARADA et al. 1990). Subsequent studies showed that the 20 kDa protein inhibits reactive lysis by blocking the activity of both C8 and C9, and the 1F5 antigen is now called homologous restriction factor of 20 kDa (HRF20).

A monoclonal antibody YTH 53.1, raised against a lymphocyte surface antigen, was shown to potentiate the susceptibility of human erythrocytes to reactive lysis (DAVIES et al. 1989). The antigen recognized by YTH 53.1 was isolated by affinity chromatography and shown by DAVIES et al. (1989) to have an Mr of approximately 20 kDa. At the 4th Leukocyte Workshop, YTH 53.1 was placed in a cluster, designated CD59, along with monoclonal antibody MEM-43. The MEM-43 antibody had also been produced by immunizing mice with membrane extracts from human lymphocytes. While it was not initially obvious that MEM-43 recognized a complement regulatory protein, subsequent data has confirmed that the MEM-43 antigen is MIRL (WHITLOW et al. 1990).

GROUX et al. (1989) developed a monoclonal antibody (H19) that blocked rosetting of human erythrocytes by lymphocytes. In addition, evidence was presented that the H19 antigen (that is also present on T lymphocytes) was required for T cell activation. While the connection with complement regulation was not made initially, it has since become evident that the 19 kDa erythrocyte antigen recognized by monoclonal antibody H19 is the same as MIRL (WHITLOW et al. 1990).

5 Relationship Between MIRL and the Erythrocyte Phenotypes of PNH

As discussed above, unlike PNH III cells, PNH II erythrocytes are resistant to CoF-initiated lysis. Either of two hypotheses seemed plausible as explanations for the molecular basis of the differences between the PNH II and the PNH III phenotypes. According to the first hypothesis, the two phenotypes have the same basic defect with the PNH III cells being more severely affected. According to the second hypothesis, PNH III erythrocytes have two independent defects. One deficiency (i.e., DAF) accounts for the intermediate complement sensitivity and the greater activation of C3, and this defect is shared by PNH II cells. The absence of a second complement regulatory protein (i.e., MIRL) is responsible for the greater susceptibility of PNH III cells to CoF-initiated hemolysis, and this factor is normal on PNH II erythrocytes. Compelling data indicate that the first hypothesis is correct (HOLGUIN et al. 1990). Using immunochemical techniques, DAF and MIRL were quantitated on PNH II and PNH III erythrocytes. As anticipated, the PNH III cells were almost completely deficient in both regulatory proteins. PNH II erythrocytes were also shown to be markedly deficient in both DAF and MIRL, but the deficiency was less severe in comparison with PNH III cells. Thus, these results indicate that the erythrocytes that are classified as PNH II have an amount of MIRL that is abnormally low but above the threshold that provides protection against reactive lysis. Quantitative analysis (by flow cytometry) of PNH II cells from 15 different patients has shown that, while the absolute amount of MIRL and DAF is variable, in each case there is a concordant deficiency of the two proteins (ROSSE et al. 1991).

In order to be fully consistent with previous observations, it is necessary to hypothesize that the deficiencies of DAF and MIRL on PNH II and PNH III cells are severe enough so that control of the amplification C3 convertase of both the classical and the alternative pathways is lost for both phenotypes (thereby accounting for the greater C3b deposition on both phenotypes), while the slightly greater amount of MIRL on PNH II cells (compared to PNH III cells) is sufficient to inhibit reactive lysis.

Inasmuch as PNH I erythrocytes are normally sensitive to complement-mediated lysis, it has been suggested that they are the progeny of residual normal hematopoietic stem cells. Quantitation of MIRL and DAF, however, suggests that, in at least some cases, PNH I cells are partially deficient in both regulatory proteins. (MEDOF et al. 1987; HOLGUIN et al. 1990). These results imply that PNH I cells may, in some instances, be derived the abnormal clone. Further, that cells that are partially deficient in DAF and MIRL have normal complement sensitivity implies that the regulatory proteins are present on normal cells in a relative excess.

6 Isolated Deficiencies of DAF and MIRL

The relative importance of DAF and MIRL in regulating susceptibility to complement-mediated lysis is brought into clear focus by cases in which there is an isolated deficiency of either one of the two inhibitors.

Recently it has been shown that antigens of the Cromer blood group complex are located on DAF, and rare cases of a null phenotype called Inab have been reported (TELEN et al. 1988). Apparently, Inab erythrocytes are completely deficient in DAF, but, unlike PNH erythrocytes, other GPI-linked proteins (including MIRL) are expressed normally by Inab cells (TELEN and GREEN 1989). As expected, Inab cells are resistant to CoF-initiated hemolysis (TELEN and GREEN 1989). While in vitro assays that depend upon activation of the classical pathway of complement have shown that Inab cells display a modest increase in susceptibility to complement-mediated lysis (TELEN and GREEN 1989), individuals with the Inab phenotype manifest no clinical evidence of hemolytic anemia (TELEN et al. 1988).

In sharp contrast to the situation of individuals with the Inab phenotype, a patient with an isolated deficiency of MIRL manifested a clinical syndrome indistinguishable from that of PNH (YAMASHINA et al. 1990). This patient experienced recurrent episodes of intravascular hemolysis; during two hemolytic crises, cerebral infarction was documented. In vitro assays confirmed that the patient's erythrocytes were abnormally sensitive to complement-mediated lysis. Together, these observations indicate that the PNH phenotype is primarily a manifestation of MIRL deficiency.

7 Relationship Between MIRL and HRF/C8bp

Currently there is no evidence that MIRL and HRF/C8bp are structurally related. The two proteins appear to be immunochemically distinct since antibodies against MIRL recognize only an 18 kDa protein when erythrocyte membrane proteins are analyzed by western blot. Conversely, using the same technique, anti-HRF and anti-C8bp do not detect an 18 kDa protein. The two proteins appear to have similar function. In agreement with observations of SUGITA et al. (1988), however, we have not observed a protein with the reported properties of HRF/C8bp in membrane extracts used to prepare MIRL. We have also demonstrated that, by blocking MIRL function with antibody, normal erythrocytes are made as susceptible to reactive lysis as PNH III erythrocytes (cell that are presumably missing HRF/C8bp) (HOLGUIN et al. 1990). Inasmuch as the anti-MIRL-treated cells would have normal HRF/C8bp, these experiments suggest that HRF/C8bp has a very limited capacity to inhibit reactive lysis. MIRL also appears to be primarily responsible for controlling susceptibilty of erythrocytes to complement-mediated hemolysis initiated by the classical pathway, because normal cells that have their MIRL function blocked manifest a complement lysis sensitivity profile that is identical to that of PNH III erythrocytes (HOLGUIN et al. 1990). Those results indicate that control of the classical pathway C3 convertase by DAF and restriction of the activity of the MAC by HRF/C8bp offer little protection against complement-mediated hemolysis if MIRL is not functional. Insight into the relationship between MIRL and HRF/C8bp will be gained when information about the primary structure of HRF/C8bp becomes available.

8 Structural Characteristics of MIRL

8.1 Primary Sequence

The primary sequence of MIRL has been determined by direct amino acid sequencing of tryptic digests (SUGITA et al. 1989a) and deduced by sequencing the cDNA from clones containing the MIRL coding region (SUGITA et al. 1989a; OKADA et al. 1989b; DAVIES et al. 1989; HOLGUIN, unpublished observation). All cDNA are predicted to encode a 128 amino acid polypeptide. Within the first 25 amino acids is contained a cluster of hydrophobic residues that is characteristic of a leader signal sequence (Fig. 4). Evidence supporting the existence of a signal peptide is provided by sequence data that has shown that the NH_2-terminal amino acid of MIRL is leucine, the 26th residue predicted from the cDNA sequence (SUGITA et al. 1989a; OKADA et al. 1989b; DAVIES et al. 1989; PARKER et al., unpublished observation). The last 28 amino acids predicted by the cDNA sequence contain a high proportion of hydrophobic residues. This finding is

MIRL cDNA Coding Sequence

1	ATGGGAATCC	AAGGAGGGTC	TGTCCTGTTC	GGGCTGCTGC	TCGTCCTGGC
51	TGTCTTCTGC	CATTCAGGTC	ATAGCCTGCA	GTGCTACAAC	TGTCCTAACC
101	CAACTGCTGA	CTGCAAAACA	GCCGTCAATT	GTTCATCTGA	TTTTGATGCG
151	TGTCTCATTA	CCAAAGCTGG	GTTACAAGTG	TATAACAAGT	GTTGGAAGTT
201	TGAGCATTGC	AATTTCAACG	ACGTCACAAC	CCGCTTGAGG	GAAAATGAGC
251	TAACGTACTA	CTGCTGCAAG	AAGGACCTGT	GTAACTTTAA	CGAACAGCTT
301	GAAAATGGTG	GGACATCCTT	ATCAGAGAAA	ACAGTTCTTC	TGCTGGTGAC
351	TCCATTTCTG	GCAGCAGCCT	GGAGCCTTCA	TCCCTAA	

Predicted Amino Acid Sequence of MIRL

1	MGIQGGSVLF	GLLLVLAVFC	HSGHSLQCYN	CPNPTADCKT	AVNCSSDFDA
51	CLITKAGLQV	YNKCWKFEHC	NFNDVTTRLR	ENELTYYCCK	KDLCNFNEQL
101	ENGGTSLSEK	TVLLLVTPFL	AAAWSLHP		

MIRL Hydrophilicity Plot

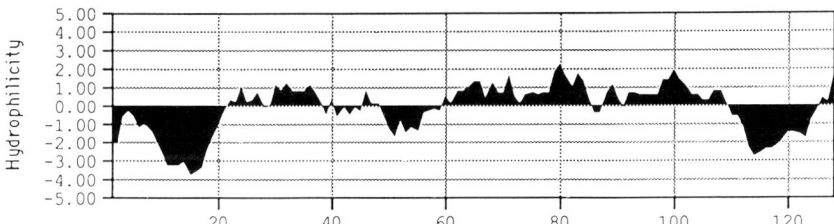

Fig. 4. Molecular characterization of MIRL. Based on the partial amino acid sequence of MIRL, mixed oligonucleotide probes were synthesized and used to screen a K562 cell cDNA library. A 1.85 kb cDNA was isolated and sequenced. Within this cDNA, an open reading frame of 387 nucleotides was identified with a nucleotide sequence predicting an amino acid sequence identical to the partial amino acid sequence of purified MIRL. This cDNA should encode a 128 amino acid peptide with the first 25 residues (underlined) resembling a typical hydrophobic leader sequence. There is a second cluster of hydrophobic residues at the COOH terminal which is presumably cleaved and replaced by a GPI anchor during posttranslational processing. (HOLGUIN et al., unpublished data)

consistent with MIRL being GPL-linked because proteins of this class are initially synthesized with a hydrophobic COOH-terminal peptide that is subsequently cleaved and replaced by a GPI anchor. (More definitive evidence that MIRL is GPI-linked is presented below.) The precise carboxyl cleavage site for GPI anchor placement has not been determined but is thought to reside between residues 66 and 85, since a tryptic fragment of the mature protein containing these residues could not be identified (SUGITA et al. 1989a).

8.2 Sequence Homology

In the case of DAF, delineation of the primary sequence provided insight into the functional properties of the molecule, since DAF was found to share a common structural motif with a group of proteins that bind to activated C3 and C4 (NICHOLSON-WELLER et al. 1986). In the case of MIRL, however, delineation of the linear sequence has proved less informative about potential structure/function relationships. Search of sequence data banks has thus far shown that MIRL shares homology only with a group of the murine lymphoid cell proteins called Ly-6 (DAVIES et al. 1989; OKADA et al. 1989b; STEFANOVA et al. 1989; SUGITA et al. 1989a). The modest degree of homology (at best 27%) between MIRL and the Ly-6 proteins is made somewhat more impressive by the fact that the positions of ten cysteines are conserved. Ly-6 is the designation for a group of GPI-linked proteins that may be involved in activation of murine T cells (PALFREE et al. 1988). There is no evidence, however, that any member of the Ly-6 family is involved in regulation of the murine complement system. Accordingly, it seems unlikely that Ly-6 is the murine analogue of MIRL. There is also a short region of homology between MIRL and DAF. Residues 6–9 of MIRL are identical to residues 129–132 of DAF (SUGITA et al. 1989b). The functional significance of this homology remains to be determined.

8.3 Glycosylation

Analysis of the primary sequence of MIRL identified two potential N-glycosylation sites (asparagine residues at positions 8 and 18 are within glycosylation signal sequences). Inasmuch as it was recovered in the appropriate yield during NH_2 terminal sequence analysis, residue 8 is apparently not glycosylated. Residue 18, however, was not detected during sequencing suggesting that it is N-glycosylated (DAVIES et al. 1989; SUGITA et al. 1989a; OKADA et al. 1989b; PARKER et al., unpublished observations). Treatment with endoglycosidase F increased the electrophoretic mobility of MIRL when analyzed by SDS-PAGE, while endoglycosidase H, neuraminidase, and endo αN-acetylgalactosaminidase treatments had no effect (SUGITA et al. 1989b; STEFANOVA et al. 1989). Together, these results indicate that MIRL has a single N-linked oligosaccharide at position 18, but apparently MIRL contains no O-linked sugars.

8.4 Sensitivity to Chemical Agents

In situ, erythrocyte MIRL is resistant to enzymatic degradation by trypsin and papain (HOLGUIN et al. 1989b; SUGITA et al. 1988, 1989b). Isolated MIRL has also been shown to be relatively resistant to both trypsin-mediated proteolysis and heat inactivation (SUGITA et al. 1989b). Analysis by immunoblotting, however, has shown that, following reduction, MIRL is no longer recognized by antibody

(HOLGUIN et al. 1989a; OKADA et al. 1989c; STEFANOVA et al. 1989). MIRL contains ten cysteine residues and apparently all are disulfide-linked. The multiple intrachain disulfide bonds apparently engender the molecule with a highly constrained secondary structure that makes it relatively resistant to proteolytic degradation and heat inactivation.

In addition to structural integrity, the functional integrity of MIRL also appears to be dependent upon intrachain disulfide bonds. In 1965, SIRCHIA et al. reported that normal human erythrocytes treated with the sulfhydryl reagent 2-aminoethylisothiouronium bromide (AET) manifested a sensitivity to complement similar to that observed for PNH erythrocytes. Subsequently, AET-treated erythrocytes have been used in studies aimed at determining the nature of the aberrant interaction of complement with PNH cells (SIRCHIA and DACIE 1967; LOGUE et al. 1973; PARKER et al. 1985). Phenotypically, AET-treated erythrocytes resemble PNH III cells in that they are susceptible to CoF-initiated hemolysis. Accordingly, we hypothesized that the sensitivity of AET-treated erythrocytes to reactive lysis was due to destruction of the structural and functional integrity of MIRL as a consequence of disruption of intrachain disulfides. Evidence in support of this hypothesis was provided by experiments that showed that, following treatment of normal erythrocytes with AET, anti-MIRL no longer bound to them (HOLGUIN et al. 1989; EZZELL et al. 1991). More extensive studies have shown that AET also completely destroys the activity of CR1, while approximately 50% of DAF function is lost following treatment of normal erythrocytes with AET (EZZELL et al. 1991). Thus, the aberrant interactions of AET-treated cells with complement are due to partial (in the case of DAF) or complete (in the case of MIRL and CR1) chemical inactivation of membrane proteins that regulate complement.

8.5 Alternative Forms of MIRL

Using radiolabeled MIRL cDNA to probe RNA from a variety of human cell lines, DAVIES et al. (1989) identified four species of RNA (0.6 kb, 1.2 kb, 1.9 kb, and 2.2 kb). The smaller two species were found in clones isolated by those investigators, but the larger two species were not. Analysis by northern blot of polyA RNA from human tonsils showed five species of MIRL RNA (6.0 kb, 2.2 kb, 1.9 kb, 1.2 kb, and 0.6 kb) (HOLGUIN et al., unpublished observations). Clones containing two of these transcripts (1.9 kb and 1.2 kb) were isolated and partially sequenced. The two clones varied in the 5' and 3' untranslated regions, but the coding regions were identical (HOLGUIN et al., unpublished data). Southern blot analysis of human genomic DNA revealed a pattern consistent with a single copy gene (DAVIES et al. 1989). Thus, at present, MIRL appears to be encoded by a single gene, and although variations in mRNA transcript size exist, sequence analysis of several cDNA has revealed only one coding sequence.

STEFANOVA et al. (1989) showed that bands representing proteins with apparent of 80 kDa and 18 kDa were immunoprecipitated from a cell line derived from peripheral blood acute lymphocytic leukemia cells by using monoclonal anti-MIRL. DAVIES et al. (1989) reported that, on some occasions, analysis by SDS-PAGE revealed a 45 kDa or an 80 kDa protein and an 18–20 kDa protein when MIRL was isolated from urine by affinity chromatography. SIMS et al. (1989) observed a band representing a protein with an apparent M_r of 37 kDa when human platelet proteins were analyzed by western blot using polyclonal anti-MIRL as the primary antibody; however, a protein with an M_r of 18–20 kDa was not visualized. The relationship between MIRL and the other proteins recognized by anti-MIRL antibodies is obscure, since sequence data is not available on any of the members of the latter group.

8.6 MIRL is GPI-Anchored

GPI-anchored proteins lack the hydrophobic membrane spanning domain that is present in the majority of integral membrane proteins. Instead, a covalent bond attaches the COOH-terminal of the protein to a glycan structure that is linked to membrane-associated phosphatidylinositol (reviewed in this volume). Proteins of this class are usually susceptible to cleavage by phosphatidylinositol-specific phospholipase C (PIPLC). This bacterial enzyme degrades phosphatidylinositol by removing 1,2 diacylglycerol. As a consequence, GPI-anchored proteins are released from the cell surface. Accordingly, susceptibility to PIPLC has been used to identify GPI-linked proteins.

As discussed above, like other GPI-linked proteins, isolated MIRL has the capacity to incorporate into erythrocytes. Accordingly, the effects of PIPLC on isolated and membrane-associated MIRL were investigated to determine if MIRL shares other characteristics with GPI-linked proteins (HOLGUIN et al. 1989b).

Normal erythrocytes were radiolabeled and incubated with either buffer or buffer containing PIPLC. Analysis by SDS-PAGE and autoradiography of proteins immunoprecipitated by anti-MIRL showed that the MIRL released into the supernatant (the PIPLC-sensitive MIRL) had an M_r of 19 kDa, whereas the MIRL that remained bound to the membrane (the PIPLC-resistant MIRL) had an M_r of 18 kDa. These studies show that cleavage by PIPLC alters the electrophoretic mobility of MIRL. While similar observations have been made for some other GPI-anchored proteins, the mechanism that accounts for this change in migration remains speculative. Further, the effects of PIPLC-mediated hydrolysis on electrophoretic mobility are inconsistent. For example, DAF and LFA-3 migrate faster after PIPLC cleavage, while the mobility of the GPI-linked form of FcγRIII appears to be unaffected (HOLGUIN et al. 1990a).

In order to quantitate the amount of MIRL released following incubation with PIPLC, a radioimmunobinding assay was used. Normal erythrocytes and K562 cells (an erythroleukemia cell line that has been shown to express MIRL) were incubated with buffer or with buffer containing PIPLC. After washing, the cells were incubated with a saturating amount of anti-MIRL, and antibody binding was subsequently quantitated by using radiolabeled anti-rabbit IgG. Treatment with PIPLC caused release of approximately 10% of the MIRL from erythrocytes and approximately 45% of MIRL from K562 cells (HOLGUIN et al. 1989b).

The effects of PIPLC on the electrophoretic mobility of isolated MIRL were also studied. Purified MIRL was incubated with buffer or with buffer containing PIPLC and subsequently analyzed by SDS-PAGE and silver stain. In contrast to its effects on a portion of membrane MIRL, PIPLC did not alter the electrophoretic mobility of isolated MIRL (HOLGUIN et al. 1990a). To determine the effects of PIPLC on the functional activity of MIRL, isolated MIRL was incubated with buffer or buffer containing PIPLC and the samples were incubated with PNH erythrocytes. After washing, susceptibility of cells to CoF-mediated lysis was determined. The inhibitory activity of PIPLC-treated MIRL was equivlalent to that of the buffer-treated control (HOLGUIN et al. 1989b). Together, these results suggest that isolated MIRL is not susceptible to hydrolysis by PIPLC.

As discussed above, PIPLC induces release of only 10% of erythrocyte MIRL. Other human erythrocyte proteins that are anchored through a GPI moiety (e.g., DAF, acetylcholinesterase, and LFA-3) demonstrate a similar resistance (DAVITZ et al. 1986; SELVARAJ et al. 1987; ROBERTS et al. 1987). Further, when isolated MIRL is treated with PIPLC, no obvious change in electrophoretic mobility is observed; after incubation with PIPLC, purified MIRL retains its capacity to reincorporate into PNH erythrocytes and to inhibit CoF-initiated hemolysis. These observations suggest that the majority of the putative inositol phosphatide associated with erythrocyte MIRL is inaccessible to PIPLC. ROBERTS et al. (1987) have reported that isolated human erythrocyte acetylcholinesterase is also resistant to PIPLC, and evidence was subsequently presented suggesting that this resistance is due to palmitoylation of the inositol ring at the 2-OH position (ROBERTS et al. 1988). Based on those observations, it has been proposed that subtle differences in the basic composition of the GPI anchor influence susceptibility to PIPLC. A preliminary report suggests that a similar mechanism accounts for the resistance of erythrocyte MIRL to PIPLC (RATNOFF et al. 1990).

Compared with erythrocyte MIRL, a greater percentage of MIRL on K562 cells is released by PIPLC. For DAF and LFA-3, susceptibility to PIPLC also varies markedly among different cell types (DAVITZ et al. 1986; SELVARAJ et al. 1987). These findings suggest that cell-specific modifications in the GPI moiety may exist for MIRL, DAF, and LFA-3 that account for the observed differences in susceptibility to PIPLC cleavage. Alternatively, susceptibility could vary if accessibility to the cleavage site of MIRL in situ were influenced by other membrane constituents. That the isolated protein is also resistant to PIPLC, however, argues against this mechanism (at least in the case of erythrocyte MIRL).

Further evidence in support of MIRL being a GPI-anchored protein was provided by experiments that showed that, when COS cells are transfected with MIRL cDNA, essentially all the MIRL that is subsequently expressed is susceptible to PIPLC (DAVIES et al. 1989). Although the proportion of susceptible molecules was variable, experiments involving a wide of variety of other cells have confirmed that MIRL is sensitive to PIPLC cleavage (STEFANOVA et al. 1989; HIDESHIMA et al. 1990; OKADA et al. 1989c; WHITLOW et al. 1990).

9 Cellular Expression of MIRL

9.1 Human Tissue

Scatchard analysis of equilibrium binding studies using radiolabeled monoclonal antibodies showed that there are 25000–30000 copies of MIRL per erythrocyte (MERI et al. 1990a; OKADA et al. 1989c). Thus, the density of MIRL is approximately ten times greater than that of DAF (KINOSHITA et al. 1985). In studies of the erythrocytes from ten normal donors, binding of anti-MIRL varied by approximately 10% suggesting that, like DAF, expression of MIRL exhibits little individual variation (HOLGUIN et al. 1989b). Immunofluorescence studies have shown that, in addition to erythrocytes, MIRL is expressed on peripheral blood lymphocytes, monocytes, granulocytes, and platelets (STEFANOVA et al. 1989; DAVIES et al. 1989; GROUX et at. 1989). Using immunohistochemical staining techniques, NOSE et al. (1990) documented that MIRL is expressed on endothelial cells of arteries, veins, and capillaries and on cultured human umbilical vein endothelial cells. In addition, MIRL was shown to be present on the Schwann sheath of peripheral nerve fibers, ependymal cells, and some epithelial cells such as bronchial epithelium, renal tubules, squamous epithelium, and acinar cells of the salivary gland.

MIRL has been affinity purified from urine with a yield of 20–100 µg/liter (DAVIES et al. 1989). The electrophoretic mobility of urine MIRL is slightly less than that of MIRL isolated from erythrocytes. That the electrophoretic mobility of MIRL released from erythrocytes by PIPLC is also slightly less than that of uncleaved MIRL suggests the urine MIRL may lack the GPI anchor, as is the case with urine DAF (MEDOF et al. 1986).

The broad tissue expression of MIRL hints at its physiologic importance. Presumably MIRL protects host cells that are in contact with serum from complement-mediated injury. The observation that MIRL is present in neural tissues was unexpected. A recent study, however, showed that MIRL expression was markely increased in tangled neurons and dystrophic neurites of Alzheimer's disease tissue in a pattern that paralleled exactly sites of MAC deposition (MCGEER et al. 1991). These results suggest that MIRL expression may be up-regulated in response to complement injury.

9.2 MIRL Expression by Cell Lines

MIRL has also been detected on the surface of a variety of cultured cell lines. K562, HL-60, MT2, Molt-4, Jurkat, CEM, HPB-ALL, KM3, HEL, ML1, EBV-transformed lymphoblastoid, and ESH92 cell lines have all been shown to express MIRL as determined by immunochemical studies (HOLGUIN et al. 1989b; STEFANOVA et al. 1989; DAVIES et al. 1989; OKADA et al. 1989c). Using the same methods, MIRL was not detected on Nalm-6, Nalm-1, Daudi, Raji, and U937 cell lines (STEFANOVA et al. 1989; DAVIES et al. 1989).

10 Analysis of the Functional Properties of MIRL

10.1 Inhibition of Complement-Mediated Lysis

Two strategies have been used to design experiments aimed at elucidating the mechanism by which MIRL inhibits complement-mediated lysis. In the first method, isolated MIRL is allowed to incorporate into the membrane of cells that are sensitive to reactive lysis, and the inhibitory activity of MIRL can then be assessed by comparison with untreated cells. In the second method, MIRL function on cells that constitutively express the protein is blocked by antibody. The effects of MIRL inhibition on complement activity can then be assessed by comparison with cells treated with nonimmune immunoglobulins.

The first method takes advantage of the fact that, when purified MIRL is incubated with cells, a portion spontaneously incorporates into the membrane, and at least some of the incorporated MIRL retains functional activity (HOLGUIN et al. 1989a). Spontaneous incorporation is characteristic of GPI-anchored proteins and presumably occurs by way of nonspecific hydrophobic interaction between the phosphatidylinositol moiety and the lipid bilayer of the cell membrane (MEDOF et al. 1984). Incubation of PNH III erythrocytes with isolated MIRL caused a dose-dependent inhibition of CoF-initiated lysis (Fig. 1) (HOLGUIN et al. 1989a). Approximately 400 molecules/cell mediated 50% inhibition, whereas approximately 1000 molecules/cell were required for complete inhibition of lysis. Similar results have been reported for MIRL inhibition of reactive lysis using isolated components of the MAC (MERI et al. 1990a; WHITLOW et al. 1990). Analysis of the binding of radiolabeled C7 and C8 to normal erythrocytes, PNH III erythrocytes, and PNH III erythrocytes treated with MIRL indicated that MIRL inhibited binding of both molecules. Inhibition of C7 binding, however, was observed only when relatively high concentrations of cells were used. The explanation for the dependence of C7 inhibition on cell concentration is speculative. Conceivably, MIRL may exert some of its inhibitory activity intercellularly rather than intracellularly, and this effect would require that the cells be in relatively close contact. Previous studies from this laboratory have also shown that, following incubation with activated CoF and EDTA serum, PNH III erythrocytes bind approximately 5 times more C5b-7 than normal erythrocytes but approximately 15 times more C9 than normal cells (PARKER et al. 1989). These results suggest that MIRL influences the binding of C9 as well as the formation of C5b-7.

Using isolated components of the MAC, MIRL has been shown to bind to C5b-8 and C5b-9 but not to C5b6 or C5b-7. (SUGITA et al. 1988; OKADA et al. 1989a; MERI et al. 1990a; WHITLOW et al. 1990). The consequence of MIRL binding to C5b-8 and C5b-9 is manifested by a reduction in C9 multiplicity. The ratio of C8:C9 binding on normal erythrocytes is approximately 1:1, whereas for PNH III cells the ratio is approximately 3:1 (PARKER et al. 1989). In agreement with those results, MERI et al. (1990a) have reported that treatment of guinea pig erythrocytes with MIRL reduced the ratio of C9:C8 binding from 3.3:1 to 1.5:1. Together, these studies suggest that MIRL inhibits complement-mediated lysis, in part by limiting the efficacy of the MAC through restriction of multiplicity of C9 binding. Further, it seems likely that the observed reduction in C9 polymerization is a direct consequence of restriction of C9 multiplicity (ROLLINS and SIMS 1990).

Although MIRL has been reported to bind to isolated C9 in the fluid phase (MERI et al. 1990a), the molecule to which MIRL binds in the C5b-8 complex has not been identified, although indirect evidence implicates that it is C8 (ROLLINS et al. 1991). Given the observation that MIRL does not bind to isolated C5b6, C7, or C8 in the fluid phase, seems likely that the MIRL binding site is generated as a result of conformational changes that occur in association with formation of the C5b-8 complex (MERI et al. 1990a). In contrast to the observed inhibition of C9 polymerization by MIRL when the MAC was generated on chicken erythrocytes using isolated human components (ROLLINS and SIMS 1990), MIRL did not inhibit zinc-induced polymerization of C9 (MERI et al. 1990a). Since the latter is a nonphysiological process, the significance of this observation remains to be determined.

In summary, MIRL appears to have the capacity to bind to both C5b-8 and C9. Binding to C5b-8 appears to inhibit C9 binding, and binding to C9 within the C5b-9 complex appears to restrict C9 multiplicity.

As anticipated, inhibition of MIRL function with antibody causes normal human erythrocytes to become susceptible to CoF-initiated lysis (Fig. 2) (HOLGUIN et al. 1989a). Others have shown that inhibition of MIRL function also induces susceptibility to reactive lysis when isolated components of the MAC are used (SUGITA et al. 1988; OKADA et al. 1989a; MERI et al. 1990a; ROLLINS and SIMS 1990).

10.2 MIRL and Nonerythroid Cells

As discussed above, MIRL is constitutively expressed by a variety of both nonerythroid cells and cultured cell lines, and MERI et al. (1990a) showed that treatment of K562 cells (an erythroleukemia cell line) with anti-MIRL induced susceptibility to CoF-initiated cytolysis. Thus, MIRL appears to function as an inhibitor of complement-mediated injury on both nucleated cells and on erythrocytes.

The observation that patients with PNH have a relatively high incidence of thromboembolic disease suggests that cellular constituents deficient in PNH may participate in regulation of the coagulation and complement systems. Accordingly, the structure and function of MIRL on platelets and endothelial cells has been investigated. SIMS et al. (1989) reported that, following treatment with polyclonal anti-MIRL, a number of procoagulant stimulatory processes were enhanced on platelets exposed to nonlytic amounts of MAC components. Making interpretation of those results difficult is the fact that analysis of platelet proteins by western blot showed that the antibody used by SIMS and colleagues bound to a protein with an M_r of 37 kDa. Presently, the relationship between this 37 kDa platelet protein and MIRL is obscure; however, when platelet proteins are analyzed in our laboratory by immunoblotting using polyclonal anti-MIRL, a 37 kDa protein is not observed (PARKER et al., unpublished observations).

More definitive data indicate that MIRL is present on cultured human umbilical vein endothelial cells. Using both polyclonal and monoclonal antibodies, HAMILTON et al. (1990) showed that endothelial cells express a protein with an M_r of 18–21 kDa. In addition, the polyclonal antibody (apparently the same antibody used by SIMS et al.) also recognized an endothelial cell protein with a larger M_r. As was the case with platelets, treatment with the polyclonal anti-MIRL enhanced some C5b-9 induced procoagulant responses by endothelial cells (e.g., stimulated secretion of von Willebrand factor and expression of catalytic surface for the prothrombinase complex).

Together, the studies of SIMS et al. (1989) and HAMILTON et al. (1990) suggest a possible mechanism by which a deficiency of MIRL may contribute to thromboembolic disease. Inasmuch as PNH is a disease of the hematopoietic stem cells, endothelial cells in PNH would be expected to express MIRL normally. This observation suggests that a deficiency of endothelial cell MIRL is not necessary to enhance the incidence of thrombosis. Thus, the increased thrombotic tendency in PNH is most likely due to the effects of aberrant regulation of complement activity by peripheral blood elements (e.g., platelets). The fact that the patient with isolated MIRL deficiency experienced episodes of cerebral infarction (YAMASHINA et al. 1990) suggests that lack of MIRL function is sufficient to induce a hypercoagulable state (i.e., absence of other GPI-linked complement regulatory proteins is not an absolute requirement). Conceivably, the putative absence of endothelial cell MIRL could potentiate the thrombotic tendency that is the cosequence of MIRL deficiency involving hematopoietic cells. Further, that the patient with the isolated deficiency of MIRL experienced both cerebral infarcts during hemolytic crises implicates complement activation in the pathophysiology of the thrombosis.

10.3 Homologous Restriction of Complement Activation

MIRL has been shown to function as a (HRF) in that it more effectively inhibits lysis mediated by human complement than that mediated by complement derived from other species. MIRL has essentially no inhibitory activity against reactive lysis mediated by rabbit or guinea pig complement (SUGITA et al. 1988;

OKADA et al. 1989a; DAVIS et al. 1989; ROLLINS et al. 1991). ROLLINS et al. (1991) also reported that MIRL has a modest inhibitory effect on sheep complement and a moderate effect on dog complement; however, MIRL inhibited baboon complement as well as or better than human complement. In that same study, evidence was presented that the differences in species restriction were dependent upon the source of both C8 and C9. For example, when human C8 was used, MIRL blocked reactive lysis regardless of the source of C9; when guinea pig C8 was used, MIRL blocked reactive lysis if the C9 was of human origin. Thus, the differences in susceptibility of human erythrocytes to complement of various sources appears to be due to properties of MIRL that determine binding to C8 and C9 in the process of assembly of the MAC.

10.4 Effects on Early Stages of Complement Activation

The standard clinical test for PNH is the acidified serum test of Ham. In this assay, test erythrocytes are incubated in serum that has been titrated to pH 6.4. Under these conditions, PNH erythrocytes are hemolyzed as a consequence of activation of the alternative pathway of complement, but normal erythrocytes are resistant. LOGUE et al. (1973) showed that, following incubation in acidified serum, PNH erythrocytes had bound much greater amounts of activated C3 than normal cells, suggesting that PNH cells are deficient in membrane constituents that regulate the activity of the amplification C3 convertase of the alternative pathway. An explanation for the aberrant effects of acidified serum on PNH cells seemed apparent when it was discovered that the erythrocytes of PNH are deficient in DAF. Studies by MEDOF et al. (1985), however, suggested that, while DAF participated in regulation of sensitivity to acidified serum lysis, a deficiency of DAF alone could not account entirely for the enhanced susceptibility of PNH erythrocytes.

We hypothesized that MIRL also participated in regulating sensitivity to acidified serum lysis. In order to assess the effects of DAF and MIRL, PNH erythrocytes were repleted with the purified proteins (WILCOX et al. 1991). DAF partially inhibited susceptibility to acidified serum by blocking C3 convertase activity. MIRL also inhibited lysis. As expected, MIRL blocked the activity of the MAC. Unexpectedly, however, MIRL also partially inhibited the activity of the alternative pathway C3 convertase. When DAF function was blocked with antibody, normal erythrocytes became partially susceptible to acidified serum lysis because regulation of C3 convertase activity was partially lost. By blocking MIRL, normal erythrocytes were made completely susceptible to lysis, and control of C3 convertase activity was partially lost. These studies indicated that MIRL has a regulatory effect on both the amplification C3 convertase of the alternative pathway and on the MAC.

To investigate the mechanism by which MIRL regulates the C3 convertase, isolated components of the convertase were used (EZZELL et al. 1991). Blocking DAF on normal erythrocytes markedly enhanced the activity of the convertase, but blocking MIRL had no effect. These results suggested that the regulatory activity of MIRL requires a serum factor that is not one of the known constituents of the alternative pathway C3 convertase. Recent studies from our laboratory indicate that there exists a serum protein that augments the functional activity of the C3 convertase and that this complement augmenting factor is inhibited by MIRL (C. J. PARKER, unpublished observations). Studies aimed at isolating and characterizing the protein are ongoing.

10.5 Effects on Cell-Mediated Lysis

Cytotoxic T lymphocytes and natural killer cells utilize a pore forming protein called perforin to induce cytolysis (reviewed in detail in this volume). Perforin shares structural and functional characteristics with components of the MAC. There is 17%–21% sequence homology between perforin and C7, C8, and C9 (LICHTENHELD et al. 1988), and antibodies developed against perforin (in a reduced state) cross-react with C7, C8, and C9 when immunoblotting experiments are performed under reducing conditions (YOUNG et al. 1986; TSCHOPP et al. 1986). In a manner analogous to the MAC, perforin forms ring-like lesions and induces changes in membrane potential. The mechanism by which cytotoxic T lymphocytes and natural killer cells resist autolysis by perforins has not been elucidated. With such prominent structural, antigenic, and functional similarities between C9 and perforin, it seemed plausible to hypothesize that MIRL might inhibit cell-mediated cytotoxicity. Recent studies by MERI et al. (1990b), however, conclusively demonstrate that MIRL does not regulate perforin-induced lysis.

10.6 MIRL and T Cell Activation

As discussed above, MIRL shares sequence homology with the products of the multigenic murine Ly-6 locus. Many of the gene products of this locus are able to initiate murine T cell activation in a non-antigen-dependent fashion by cross-linking the Ly-6 antigen with antibody (MALEK et al. 1986; YEH et al. 1987). Further, some Ly-6 proteins have been reported to enhance antigen-dependent T cell functions (ROCK et al. 1986). Studies of Ly-6 deficient mutant cell lines and mutant cell lines deficient in GPI-anchored proteins have demonstrated impaired immune responses (YEH et al. 1988), suggesting a physiologically significant role for these molecules in T cell activation. A number of other GPI anchored proteins have also been reported to modulate non-antigen-dependent T cell activation after cross-linking with antibody, raising the possibility that the GPI anchor may initiate the transduction of an activating signal (REISER et al. 1986). (A detailed discussion of GPI anchored proteins and T cell activation is included in this volume.)

Since MIRL is a GPI-anchored protein that shares sequence homology with murine T cell activating molecules, the effects of MIRL on T cell activation have been investigated. HIDESHIMA et al. (1990) demonstrated that MIRL was present on human T cells and that, after cross-linking it with antibody, thymidine uptake was enhanced. GROUX et al. (1989) developed a monoclonal antibody (H19) that bound to human erythrocytes and blocked T cell rosetting. Subsequent studies demonstrated that MIRL is the H19 antigen (WHITLOW et al. 1990). When peripheral blood mononuclear cells were incubated with the monoclonal anti-MIRL, thymidine uptake following CD3 stimulation was blocked. This inhibitory effect was specific for CD3 stimulation inasmuch as stimulation by lectins, allogeneic cells, or mitogenic pairs of CD2 antibodies was unaffected. Additional experiments indicated that anti-MIRL binding to antigen presenting cells, rather than to T cells, was responsible for the inhibitory activity. The observation that anti-MIRL blocks T cell rosetting suggests the possibility that T cells express a specific receptor for MIRL. Conceivably, interactions between monocyte MIRL and the T cell receptor for MIRL lead to enhanced CD3-dependent T cell activation. Whether complement activation modulates the effects of MIRL on T cell function remains to be determined (WHITLOW et al. 1990).

11 Conclusions

The greater susceptibility of PNH erythrocytes to complement-mediated lysis is primarily a manifestation of MIRL deficiency. MIRL inhibits complement-induced hemolysis by blocking the assembly of the MAC. Compelling evidence indicates that MIRL binds to C5b-8 and inhibits C9 binding and that MIRL binds to C5b-9 and restricts C9 multiplicity. In the process of acidified serum lysis, MIRL also appears to modulate the activity of the alternative pathway C3 convertase, although the mechanism of regulation has not been elucidated.

Future studies will be aimed at identifying the C8 and C9 binding sites on MIRL, at establishing the molecular basis of C3 convertase regulation, and at determining the function of MIRL in nonerythroid cells. In vitro studies suggest that MIRL activity is not restricted to the complement system; however, the physiological significance of these observations is speculative.

References

Davies A, Simmons DL, Hale G, Harrison RA, Tighe H, Lachmann PJ, Waldmann H (1989) CD59, an Ly-6-like protein expressed in human lymphoid cells, regulates the action of the complement membrane attack complex on homologous cells. J Exp Med 170: 637 654

Davitz MA, Low MG, Nussenzweig V (1986) Release of decay-accelerating factor (DAF) from the cell membrane by phosphatidylinositol-specific phospholipase C (PIPLC): selective modification of a complement regulatory protein. J Exp Med 163: 1150–1161

Ezzell JL, Wilcox LA, Bernshaw NJ, Parker CJ (1991) Induction of the paroxysmal nocturnal hemoglubinuria phenotype in normal human erythrocytes. Blood (in press)

Groux H, Huet S, Aubrit F, Tran HC, Boumsell L, Bernard A (1989) A 19-kDa human erythrocyte molecule H19 is involved in rosettes, present on nucleated cells, and required for T cell activation. J Immunol 142: 3013–3020

Hamilton KK, Ji A, Rollins S, Stewart BH, Sims PJ (1990) Regulatory control of the terminal complement proteins at the surface of human endothelial cells: neutralization of a C5b-9 inhibitor by antibody to CD59. Blood 76: 2572–2577

Hänsch GM, Hammer CH, Vanguri P, Shin ML (1981) Homologous species restriction in lysis of erythrocytes by terminal complement proteins. Proc Natl Acad Sci USA 78: 5118–5121

Hänsch GM, Schönermark S, Roelcke D (1987) Paroxysmal nocturnal hemoglobinuria type III. Lack of an erythrocyte membrane protein restricting the lysis by C5b-9. J Clin Invest 80: 7–12

Harada R, Okada N, Fujita T, Okada H (1990) Purification of 1F5 antigen that prevents complement attack on homologous cell membranes. J Immunol 144: 1823–1828

Hideshima T, Okada N, Okada H (1990) Expression of HRF20, a regulatory molecule of complement activation, on peripheral blood mononuclear cells. Immunology 69: 396–401

Holguin MH, Fredrick LR, Bernshaw NJ, Wilcox LA, Parker CJ (1989a) Isolation and characterization of a membrane protein from normal human erythrocytes that inhibits reactive lysis of the erythrocytes of paroxysmal nocturnal hemoglobinuria. J Clin Invest 84: 7–17

Holguin MH, Wilcox LA, Bernshaw NJ, Rosse WF, Parker CJ (1989b) Erythrocyte membrane inhibitor of reactive lysis: effects of phosphatidylinositol-specific phospholipase C on the isolated and cell-associated protein. Blood 75: 284–289

Holguin MH, Wilcox LA, Bernshaw NJ, Rosse WF, Parker CJ (1990) Relationship between the membrane inhibitor of reactive lysis and the erythrocyte phenotypes of paroxysmal nocturnal hemoglobinuria. J Clin Invest 84: 1387–1394

Hu V, Shin ML (1984) Species-restricted target cell lysis by human complement: complement-lysed erythrocytes from heterologous and homologous species differ in their ratio of bound to inserted C9. J Immunol 133: 2133–2137

Kinoshita T, Medof ME, Silber R, Nussenzweig V (1985) Distribution of decay-accelerating factor in the peripheral blood of normal individuals and patients and paroxysmal nocturnal hemoglobinuria. J Exp Med 162: 75–92

Lachmann PJ, Halbwachs L (1975) The influence of C3b inactivator (KAF) concentration on the ability of serum to support complement activation. Clin Exp Immunol 21: 109–114

Lichtenheld MG, Olsen KJ, Lu P, Lowrey D, Hameed A, Hengartner H, Podack ER (1988) Structure and function of human perforin. Nature 335: 448–451

Logue GL, Rosse WF, Adams JP (1973) Mechanisms of immune lysis of red blood cells in vitro. I. Paroxysmal nocturnal hemoglobinuria erythrocytes. J Clin Invest 52: 1129–1137

Malek TR, Ortega G, Chan C, Kroczek R, Shevach EM (1986) Role of Ly-6 in T cell activation. III. Induction of T cell activation by monoclonal anti-Ly-6 antibodies. J Exp Med 164: 709–722

McGeer PL, Walker DG, Akiyama H, Kawamata T, Guan AL, Parker CJ, Okada N, McGeer EG (1991) Detection of the membrane inhibitor of reactive lysis (CD59) in diseased neurons of Alzheimer brain. Brain Res (in press)

Medof ME, Kinoshita T, Nussenzweig V (1984) Inhibition of complement activation on the surface of cells after incorporation of decay-accelerating factor (DAF) into their membranes. J Exp Med 160: 1558–1578

Medof ME, Kinoshita T, Silber R, Nussenzweig V (1985) Amelioration of lytic abnormalities of paroxysmal nocturnal hemoglobinuria with decay-accelerating factor. Proc Natl Acad Sci USA 82: 2980–2984

Medof ME, Walter EI, Rutgers JL, Knowles DM, Nussenzweig V (1986) Identification of the complement decay-accelerating factor (DAF) on epithelium and glandular cells and in body fluids. J Exp Med 165: 848–864

Medof ME, Gottlieb A, Kinoshita T, Hall S, Silber R, Nussenzweig V, Rosse WF (1987) Relationship between decay accelerating factor deficiency, acetylcholinesterase activity, and defective terminal complement pathway restriction in paroxysmal nocturnal hemoglobinuria erythrocytes. J Clin Invest 80: 165–174

Meri S, Morgan BP, Davies A, Daniels RH, Olavesen MG, Waldmann H, Lachmann PJ (1990a) Human protectin (CD59), an 18-20 kD complement lysis restricting factor, inhibits C5b-8 catalyzed insertion of C9 into lipid bilayers. Immunology 72: 1–9

Meri S, Morgan BP, Wing M, Jones J, Davies A, Podack E, Lachmann PJ (1990b) Human protectin (CD59), an 18–20-KD homologous complement restriction factor, does not restrict perforin-mediated lysis. J Exp Med 172: 367–370

Nicholson-Weller A, Zaia J, Raum MG, Coligan JE (1986) Decay accelerating factor (DAF) peptide sequences share homology with a consensus sequence found in the superfamily of structurally related complement proteins including haptoglobin, factor XIII, B2-glycoprotein I, and the IL-2 receptor. Immunol Lett 14: 307–311

Nose M, Katoh M, Okada N, Kyogoku M, Okada H (1990) Tissue distribution of HRF20, a novel factor preventing the membrane attack of homologous complement, and its predominant expression on endothelial cells in vivo. Immunology 70: 145–149

Okada N, Harada R, Fujita T, Okada H (1989a) A novel membrane glycoprotein capable of inhibiting membrane attack by homologous complement. Int Immunol 1: 205–208

Okada H, Nagami Y, Takahashi K, Okada N, Hideshima T, Takizawa H, Kondo J (1989b) 20 kDa homologous restriction factor of complement resembles T cell activating protein. Biochem Biophys Res Commun 162: 1553–1559

Okada N, Harada R, Fujita T, Okada H (1989c) Monoclonal antibodies capable of causing hemolysis of neuraminidase-treated human erythrocytes by homologous complement. J Immunol 143: 2262–2266

Packman CH, Rosenfeld SI, Jenkins DE, Thiem PA, Leddy JP (1979) Complement lysis of human erythrocytes: differing susceptibility of 2 types of paroxysmal nocturnal hemoglobinuria cells to C5b-9. J Clin Invest 64: 428–433

Palfree RGE, Sirlin S, Dumont FJ, Hämmerling U (1988) N-terminal and cDNA characterization of murine lymphocyte antigen Ly-6C.2. J Immunol 140: 305–310

Parker CJ, Wiedmer T, Sims PJ, Rosse WF (1985) Characterization of the complement sensitivity of paroxysmal nocturnal hemoglobinuria erythrocytes. J Clin Invest 75: 2074–2084

Parker CJ, Stone OL, Bernshaw NJ (1989) Characterization of the enhanced susceptibility of paroxysmal nocturnal hemoglobinuria erythrocytes to complement-mediated hemolysis initiated by cobra venom factor. J Immunol 142: 208–216

Rathoff WD, Knez JJ, Prince GM, Medof ME (1990) Structural properties of the glycoplasmanylinositol anchor structure of the complement membrane attack complex inhibitor CD59. Blood 76: 389a

Reiser H, Oettgen H, Yeh ETH, Terhorst C, Low MG, Benacerraf B, Rock KL (1986) Structural characterization of the TAP molecule: a phosphatidylinositol-linked glycoprotein distinct from the T cell receptor/T3 complex and Thy-1. Cell 47: 365–370

Roberts WL, Kim BH, Rosenberry TL (1987) Differences in the glycolipid membrane anchors of bovine and human erythrocyte acetylcholinesterases. Proc Natl Acad Sci USA 84: 7817–7821

Roberts WL, Myher JJ, Kuksis A, Low MG, Rosenberry TL (1988) Lipid analysis of the glycoinositol phospholipid membrane anchor of human erythrocyte acetylcholinesterase. J Biol Chem 263: 18766–18775

Rock KL, Yeh ETH, Gramm CF, Haber SI, Reiser H, Benacerraf B (1986) TAP, a novel T cell-activating protein involved in the stimulation of MHC-restricted T lymphocytes. J Exp Med 163: 315–333

Rollins SA, Sims PJ (1990) The complement-inhibitory activity of CD59 resides in its capacity to block incorporation of C9 into membrane C5b-9. J Immunol 144: 3478–3483

Rollins SA, Zhao JI, Ninomiya H, Sims PJ (1991) Inhibition of homologous complement by CD59 is mediated by a species-selective recognition conferred through binding to C8 within C5b-89 or C9 within C5b-9. J Immunol 146: 2345–2351

Rosenfeld SI, Packman CH, Jenkins DE, Countryman JK, Leddy JP (1980) Complement lysis of human erythrocytes III. Differing effectiveness of human and guinea pig C9 on normal and paroxysmal nocturnal hemoglobinuria cells. J Immunol 125: 2063–2068

Rosse WF, Parker CJ (1985) Paroxysmal nocturnal hemoglobinuria. Clin Haematol 14: 105–125

Rosse WF, Hall S, Campbell M, Borowitz M, Moore JO, Parker CJ (1991) The erythrocytes in paroxysmal nocturnal hemoglobinuria of intermediate sensitivity to complement lysis (to be published)

Schönermark S, Rauterberg EW, Shin ML, Loke S, Roelcke D, Hänsch GM (1986) Homologous species restriction in lysis of human erythrocytes: a membrane-derived protein with C8-binding capacity functions as an inhibitor. J Immunol 136: 1772–1776

Selvaraj P, Dustin ML, Silber R, Low MG, Springer TA (1987) Deficiency of lymphocyte function-associated antigen 3 (LFA-3) in paroxysmal nocturnal hemoglobinuria. J Exp Med 166: 1011–1025

Sims PJ, Rollins SA, Wiedmer T (1989) Regulatory control of complement on blood platelets. Modulation of platelet procoagulant responses by a membrane inhibitor of the C5b-9 complex. J Biol Chem 264: 19228–19235

Sirchia G, Dacie JV (1967) Immune lysis of AET-treated normal red cells (PNH-like cells). Nature 215: 747–748

Sirchia G, Ferrone S, Mercuriali F (1965) The action of two sulfhydryl compounds on normal human red cells: relationship to red cells of paroxysmal nocturnal hemoglobinuria. Blood 25: 502–509

Stefanova I, Hilgert I, Kristofova H, Brown R, Low MG, Horejsi V (1989) Characterization of a broadly expressed human leucocyte surface antigen MEM-43 anchored in membrane through phosphatidylinositol. Mol Immunol 26: 153–161

Sugita Y, Nakano Y, Tomita M (1988) Isolation from human erythrocytes of a new membrane protein which inhibits the formation of complement transmembrane channels. J Biochem 104: 633–637

Sugita Y, Tobe T, Oda E, Tomita M, Yasukawa K, Yamaji N, Takemoto T, Furuichi K, Takayama M, Yano S (1989a) Molecular cloning and characterization of MACIF, an inhibitor of membrane channel formation of complement. J Biochem 106: 555–557

Sugita Y, Mazda T, Tomita M (1989b) Amino-terminal amino acid sequence and chemical and functional properties of a membrane attack complex-inhibitory factor from human erythrocyte membranes. J Biochem 106: 589–592

Telen MJ, Green AM (1989) The INAB phenotype: characterization of the membrane protein and complement regulatory defect. Blood 74: 437–441

Telen MJ, Hall SE, Green AM, Moulds JJ, Rosse WF (1988) Identification of human erythrocyte blood group antigens on decay-accelerating factor (DAF) and an erythrocyte phenotype negative for DAF. J Exp Med 167: 1993–1998

Tschopp J, Masson D, Stanley KK (1986) Structural/functional similarity between proteins involved in complement- and cytotoxic T-lymphocyte-mediated cytolysis. Nature 322: 831–834

Vogel C-W, Müller-Eberhard HJ (1982) The cobra venom factor-dependent C3 convertase of human complement. J Biol Chem 257: 8292–8299

Vogel C-W, Smith CA, Müller-Eberhard HJ (1984) Cobra venom factor: structural homology with the third component of human complement. J Immunol 133: 3235–3241

Whitlow MB, Iida K, Stefanova I, Bernard A, Nussenzweig V (1990) H19, a surface membrane molecule involved in T-cell activation, inhibits channel formation by human complement. Cell Immunol 126: 176–184

Wilcox LA, Ezzell JL, Bernshaw NJ, Parker CJ (1991) Molecular basis of the enhanced susceptibility of the erythrocytes of paroxysmal nocturnal hemoglobinuria to hemolysis in acidified serum. Blood (in press)

Yamamoto K (1977) Lytic activity of C5-9 complexes for erythrocytes from the species other than sheep: C9 rather than C8-dependent variation in lytic activity. J Immunol 119: 1482–1485

Yamashina M, Ueda E, Kinoshita T, Takami T, Ojima A, Ono H, Tanaka H, Kondo N, Orii T, Okada N, Okada H, Inoue K, Kitani T (1990) Inherited complete deficiency of 20-kilodalton homologous restriction factor (CD59) as a cause of paroxysmal nocturnal hemoglobinuria. N Engl J Med 323: 1184–1189

Yeh ETH, Reiser H, Daley J, Rock KL (1987) Stimulation of T cells via the TAP molecule, a member in a family of activating proteins encoded in the Ly-6 locus. J Immunol 138: 91–97

Yeh ETH, Reiser H, Bamezai A, Rock KL (1988) TAP transcription and phosphatidylinositol linkage mutants are defective in activation through the T cell receptor. Cell 52: 665–674

Young JD-E, Cohn ZA, Podack ER (1986) The ninth component of human complement and the pore-forming protein perforin 1 from cytotoxic T cells: structural, immunological and functional similarities. Science 233: 184–190

Zalman LS, Wood LM, Müller-Eberhard HJ (1986) Isolation of a human erythrocyte membrane protein capable of inhibiting expression of homologous complement transmembrane channels. Proc Natl Acad Sci USA 83: 6975–6979

Zalman LS, Wood LM, Frank MM, Müller-Eberhard HJ (1987) Deficiency of the homologous restriction factor in paroxysmal nocturnal hemoglobinuria. J Exp Med 165: 572–577

Homologous Restriction Factor

L. S. ZALMAN

1 Introduction	87
2 Homologous Restriction	88
3 Characterization of HRF	89
4 Paroxysmal Nocturnal Hemoglobinuria	89
5 Soluble HRF	91
6 Relationship to Other Complement Proteins	93
7 Presence in Cells Other than Erythrocytes	94
8 Inhibition of Antibody-Dependent Cellular Cytotoxicity	94
9 Cell Surface HRF in Cytotoxic Lymphocytes	95
10 HRF from Cytotoxic Granules	95
11 Protection Against Cell-Mediated Cytotoxicity	96
12 Conclusion	97
References	97

1 Introduction

It is known that human complement is more efficient at lysing erythrocytes of other species than at lysing homologous erythrocytes. This phenomenon has been called homologous species restriction (HÄNSCH et al. 1981) of complement-mediated hemolysis. Human complement is restricted by at least three proteins known so far. Decay accelerating factor (NICHOLSON-WELLER et al. 1982) is a powerful regulator of cell bound C3/C5 convertase. At the end of the complement cascade, two separate proteins seem to act. The lower molecular weight protein was first isolated as P18 (SUGITA et al. 1988) and is also known as IF5 or HRF20 (OKADA et al. 1989), MIRL (HOLGUIN et al. 1989), and CD59 (DAVIES et al. 1989). The other higher molecular weight protein is homologous restriction factor (ZALMAN et al. 1986a), also known as C8 binding protein (SCHÖNERMARK

Scripps Clinic and Research Foundation, IMM-19, 10666 North Torrey Pines Road, La Jolla, CA 92037, USA

et al. 1986) and MIP (WATTS et al. 1990). This review will focus on homologous restriction factor.

2 Homologous Restriction

YAMAMOTO (1977), using the reactive lysis system which requires only C5b6, C7, C8, and C9 (LACHMANN and THOMPSON 1970), found that the species of the C9 source was critical for lysis of erythrocytes. Guinea pig erythrocytes with human C5b6, C7, and C8 failed to lyse with the addition of guinea pig C9 but lysed very well in the presence of goat or sheep C9. Furthermore, the species of the C8 source was not as important as that of the C9. When guinea pig erythrocytes were treated with C5b6^{hu} and C7^{hu} and either C8^{gp} or C8^{hu} followed by guinea pig C9, the amount of lysis was low in both cases, although the amount with C8^{hu} was slightly greater. In a similar study by HÄNSCH et al. (1981), the amount of lysis of erythrocytes from various species by C5b6, C7, and serum, as a source of C8 and C9, was determined. In each case, when the erythrocytes and the serum were from the same species, the amount of lysis was very low. An experiment was done with human erythrocytes to assess the relative importance of C8 and C9 in this system. Human erythrocytes with C5b67^{hu} (EC5b67^{hu} treated with buffer lysed about eight times better with C8 and C9 from rabbit than did EC5b67^{hu}) that had been preincubated with human C8 and C9. Preincubation with human C8 alone reduced the amount of lysis only slightly (HÄNSCH et al. 1981). These results suggest that C9 plays the major role in the species specificity while C8 has a lesser effect. This phenomenon was termed "homologous species restriction" of complement-mediated hemolysis (HÄNSCH et al. 1981).

The protein responsible for this homologous species restriction was identified as C8 binding protein (C8bp) (SCHÖNERMARK et al. 1986) and as homologous restriction factor (HRF) (ZALMAN et al. 1986a). This 65 kDa protein

Table 1. Effect of anti-HRF on lysis of E_H C5b-7 by C8 and C9 from different species.

Species of C8 and C9	Percent lysis with anti-HRF
	Percent lysis with anti-DAF
Human	20.0
Spider Monkey	0.9
Sheep	1.1
Mouse	1.1
Donkey	0.9
Emu	0.8
Rabbit	1.0
Goat	1.0
Rat	1.0

was localized in protein blots of whole human erythrocyte membranes that were probed first with C8 and then with anti-C8. HRF was isolated by passing detergent-solubilized human erythrocyte membranes over a column which had 6 mg of human C9 bound to it. The active material was eluted from the column with a high salt buffer in detergent. This protein was incorporated into liposomes and, as such, was able to inhibit the channel-forming ability of human complement C5b-9. Pretreatment of human erythrocytes with anti-HRF caused a 20-fold increase in C5b-9 lysis due to human C8 and C9. When nonhuman C8 and C9 were added to HRF antibody-treated human EC5b-7 cells, the amount of lysis was equivalent to human erythrocytes treated with anti-DAF; that is, there was no enhancement. HRF antibody treatment of erythrocytes only affected lysis via homologous complement components. For this reason, the protein was called homologous restriction factor (ZALMAN et al. 1986a) (Table 1).

3 Characterization of HRF

In order to examine the functional characteristics of HRF, it was isolated from erythrocyte membranes. This membrane-derived HRF was incorporated into phosphotidylcholine liposomes. The rate of liposome swelling caused by C5b-9 addition, corresponding to channel formation, could be inhibited up to about 90% by the presence of HRF in the liposome membrane. HRF was also able to inhibit the channel-forming function of both polyC9 and C5b-8. HRF, then, is able to interact with both of the terminal complement components C8 and C9. In order to determine the effect of HRF on the uptake of complement components, human erythrocytes were treated with anti-HRF to block the action of HRF on the membrane. Although anti-HRF caused a 20-fold increase in the amount of reactive lysis of these erythrocytes, the uptake of radiolabeled C7 did not increase. Anti-HRF treatment did cause a fivefold increase in C9 binding, showing that HRF did not affect the number of C5b-7 sites on the membranes but did affect the number of C9 molecules that were able to bind at each site (ZALMAN et al. 1986a). HRF is not only able to bind to C8 and C9, it also shares an immunological relationship to these proteins. Antibodies to C8 and C9 detect HRF in dot blots (ZALMAN et al. 1986a) and in western blots (SCHÖNERMARK et al. 1988).

4 Paroxysmal Nocturnal Hemoglobinuria

Studies on the erythrocytes of patients suffering from the disease paroxysmal nocturnal hemoglobinuria (PNH) revealed that these cells are abnormally lytic in response to complement (ROSSE 1973; PACKMAN et al. 1979). PNH cells have been

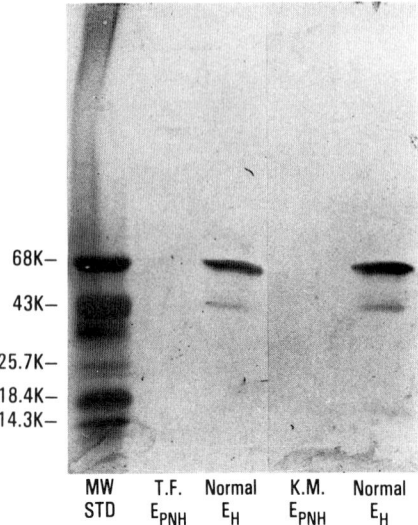

Fig. 1. Anti-HRF immunoblot of PNH and normal erythrocyte membranes. Samples of 50 µg of total protein from erythrocyte membranes of PNH patient TF, PNH patient KM, and of two normal individuals were subjected to SDS-PAGE. The proteins were transferred onto nitrocellulose and probed with rabbit anti-HRF. (From ZALMAN et al. 1987a)

shown to be deficient in decay accelerating factor (DAF) (NICHOLSON-WELLER et al. 1983; PANGBURN et al. 1983a). DAF deficiency can explain the markedly extended half-life of C3 convertase on the surface of PNH erythrocytes (PANGBURN et al. 1983a, b) and, in part, the susceptibilities of these cells to lysis in acidified serum (DACIE and RICHARDSON 1943). However, DAF deficiency cannot account for the increased sensitivities of PNH erythrocytes to reactive lysis by C5b-9 (ROSENFELD et al. 1985; SHIN et al. 1986). Additionally, the amounts of both C9 binding and C9 polymerization have been shown to be higher than normal on PNH erythrocytes (HU and NICHOLSON-WELLER 1985). Indeed, it was found that PNH erythrocytes were missing HRF (ZALMAN et al. 1987a). DAF-deficient PNH erythrocyte (PNH-E) membranes were compared to normal erythrocyte membranes by SDS-PAGE followed by western blotting with anti-HRF (Fig. 1) (ZALMAN et al. 1987a). The PNH-E membranes were clearly deficient in HRF. HRF isolated from human erythrocytes could be inserted into these PNH-E membranes by incubating the purified protein with the membranes for 2 h at 37°C. The HRF-treated PNH-E were then less susceptible to reactive lysis by C5b-9. Interestingly, only about 1000 molecules of HRF per cell were needed to reduce their susceptibility to C5b-9 to approximately normal levels, indicating that this is probably close to the normal amount of HRF in a normal erythrocyte.

When PNH-E were incubated with C8-deficient human serum, then with human C8 and C9, a large amount of polyC9 could be distinguished on SDS-PAGE. When normal human erythrocytes were treated in the same manner, the polyC9 band was missing. However, when PNH-E (containing C1–7 as before) were incubated with HRF and then human C8 and C9, the formation of polyC9 was inhibited (HÄNSCH et al. 1987). These results indicate that HRF prevents polyC9 formation on normal erythrocytes.

5 Soluble HRF

The HRF that has been described so far is a membrane constituent isolated from erythrocytes and kept in a low concentration of detergent. Recently, a soluble form of HRF was described that was isolated from human urine (ZALMAN et al. 1989). The soluble HRF derived from urine (HRF-U) had the same molecular weight and reacted identically on a western blot with polyclonal anti-HRF (Fig. 2). HRF-U, in contrast to membrane-derived HRF (HRF-M), does not spontaneously reinsert into erythrocyte membranes (ZALMAN et al. 1989; WATTS et al. 1990). However, it is fully able to inhibit reactive lysis of chicken erythrocytes (Fig. 3). This is similar to the case with DAF, as the urine form of DAF is unable to reinsert into membranes (MEDOF et al. 1987). HRF-U was used to investigate the immunobiological relatedness of HRF and other proteins of the complement system. Antibody to HRF cross-reacted most strongly with C8 and to a lesser extent with C7 and C9 (ZALMAN et al. 1989). HRF shows some homology with most (if not all) of the terminal complement components.

Interestingly, although HRF-U was able to inhibit reactive lysis, it did so in a different way from HRF-M. When HRF-M was incorporated into an erythrocyte membrane, it had no effect on the number of radioactive C7 molecules that were

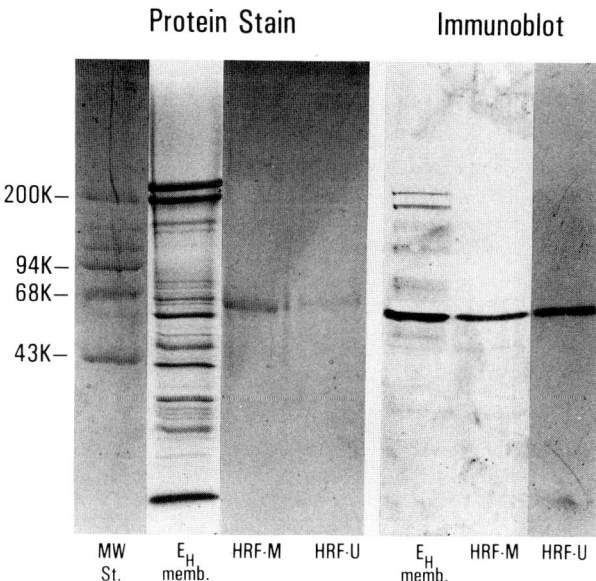

Fig. 2. Comparative analysis of HRF-U and HRF-M by SDS-PAGE and immunoblotting by using anti-HRF-M. The amount of protein applied was 3 µg HRF-M and 2 µg HRF-U. For comparison 35 µg of solubilized human erythrocyte membrane proteins (E_H memb.) were analyzed. The gel concentration gradient was 1.7%–17%, the protein was stained with Coomassie blue, and the immunoblot was developed with anti-HRF-M. (From ZALMAN et al. 1989).

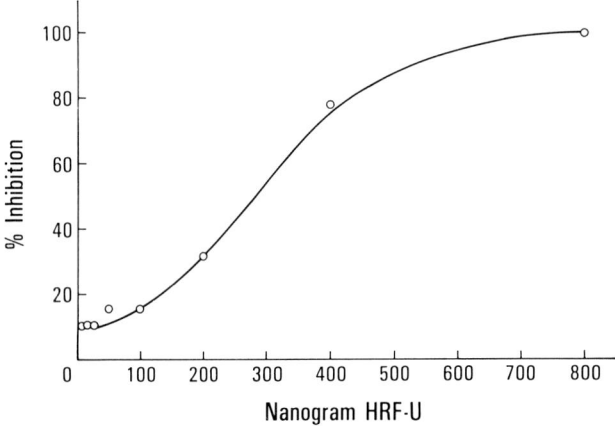

Fig. 3. Inhibition of reactive lysis of E_c by HRF-U. E_c (10^7) were incubated with 0–800 ng of purified HRF-U for 10 min at 37°C. The cells were then subjected to reactive lysis with the addition of human C5b6, C7, C8, and C9 for 20 min at 37°C. Lysis in the absence of HRF-U was 89%. (From ZALMAN et al. 1989)

able to bind to the cell as part of a C5b-7 complex. Anti-HRF had no effect on the uptake of radiolabeled C7 on the membrane (ZALMAN et al. 1986a). It also had no effect on ^{125}I-C8 uptake on the membrane (ZALMAN and MÜLLER-EBERHARD, in preparation). HRF-M, inserted into an erythrocyte membrane, had a substantial negative effect on C9 binding (ZALMAN and MÜLLER-EBERHARD, in preparation) and on C9 polymerization (SCHÖNERMARK et al. 1988). HRF-M, then, interacts with C8 and/or C9 on the cell membrane to prevent binding and polymerization of C9 (SCHÖNERMARK et al. 1988). The actual site of HRF-M action is not yet clear. One of the ways that HRF was first detected is by its affinity for C8 (SCHÖNERMARK et al. 1986). The affinity of C8 for radiolabeled HRF could also be demonstrated by a shift in molecular weight in an ultracentrifugation study "in low ionic strength." To further investigate the binding of HRF to C8, C8 containing α, β, and γ subunits was separated by SDS-PAGE under reducing and nonreducing conditions. Under reducing conditions, HRF bound to the α–γ subunit and under nonreducing conditions only to the γ-chain (SCHÖNERMARK et al. 1988). However, when a form of C8 lacking the γ chain was used to lyse erythrocytes, the amount of C9 binding was the same as with normal C8. Thus, the presence or absence of the γ-chain made no difference in C8 binding to EAC1-8 (DAVÉ and SODETZ 1990). The role of the C8 subunits and which of these in particular interacts with HRF remains unclear.

HRF-U, on the other hand, seems not to interact with C8 or C9. In order to effect inhibition of reactive lysis, HRF-U had to be introduced into the reaction mixture before the addition of C7. It was most effective when added prior to C5b6 (Fig. 4). When HRF-U was added after C7, the inhibition of C5b-9 lysis was minimal. These results indicate that the fluid phase HRF-U acts at the C5b-7

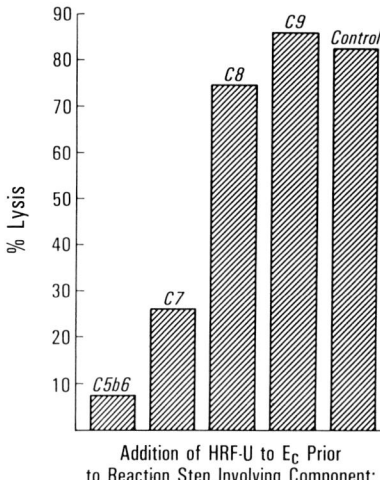

Fig. 4. Reaction steps of C5b-9 lysis affected by HRF-U. The sequence of treatment of 10^7 E_c was as follows: (a) HRF-U (10 min, 37°C), C5b6, C7, C8, C9; (b) C5b6, HRF-U (10 min, 37°C), C7, C8, C9; (c) C5b6, C7, HRF-U (10 min, 37°C), C8, C9; (d) C5b6, C7, C8, HRF-H (10 min, 37°C), C9; (e) C5b6, C7, C8, C9 (control). The amounts used were: HRF-U, 800 ng; C5b6, 1.5 µg; C7, 1 µg; C8, 1 µg; C9, 5 µg. All samples were incubated at 37°C for the same period of time. (From ZALMAN et al. 1989)

reaction step and has little effect on C8 and C9 action, in contrast to the membrane form which acts at the C8 or C9 step of reactive lysis (ZALMAN et al. 1989). In theory, HRF should be able to interfere with each step of membrane attack complex formation since C6, C7, C8, and C9 are homologous proteins. However, metastable C5b-7 should be more accessible to the fluid phase form of HRF, whereas membrane insertion of C8 and C9 should be more readily controlled by HRF-M.

6 Relationship to Other Complement Proteins

HRF is one of a series of proteins that is missing in the disease PNH. The others are DAF, CD59, and acetylcholin-esterase. Thus, a relationship between these proteins was sought. Although there is a report that HRF has a glycan phosphatidylinositol (GPI) anchor (HÄNSCH et al. 1988) like DAF (DAVITZ et al. 1986), no immunological relationship between the two has been found. HRF has no effect on the stability of C3 convertase (SCHÖNERMARK et al. 1986), and anti-DAF did not enhance the species-restricted lysis of the late complement components (ZALMAN et al. 1986a; SHIN et al. 1986). HRF is antigenically different from acetylcholin-esterase (HÄNSCH 1988) and is probably most similar to CD59. The function of these proteins is similar in that they both block reactive lysis.

A relationship between HRF and S protein was sought because they both interfere with C5b-7 attachment to cells and inhibit C9 polymerization (PODACK et al. 1984). However, antibodies to HRF do not react to S protein nor do antibodies to S protein react to HRF (ZALMAN et al. 1989; WATTS et al. 1990). HRF, though, is related to the terminal complement components. HRF-U was analyzed with antisera to C7, C8, and C9 by ELISA. Anti-HRF cross-reacted most strongly with C8 and to a lesser extent with C7 and C9.

7 Presence in Cells Other than Erythrocytes

Using antibodies to HRF, the presence of this protein on cells other than erythrocytes was explored. HRF was found on the surface of PMNs (ZALMAN et al. 1986a) and several other blood cells such as monocytes. Similar to the case with erythrocytes, the addition of anti-HRF to monocytes renders them more susceptible to reactive lysis (Fig. 5), again showing the importance of HRF in protecting cells from homologous complement.

8 Inhibition of Antibody-Dependent Cellular Cytotoxicity

Due to certain structural and functional similarities between C9 and perforin (C9RP) (ZALMAN et al. 1986b), the cytolytic protein of human killer lymphocytes, HRF was tested to see whether it might affect the function of the cellular killing process. The antibody-dependent cellular cytotoxicity (ADCC) reaction was

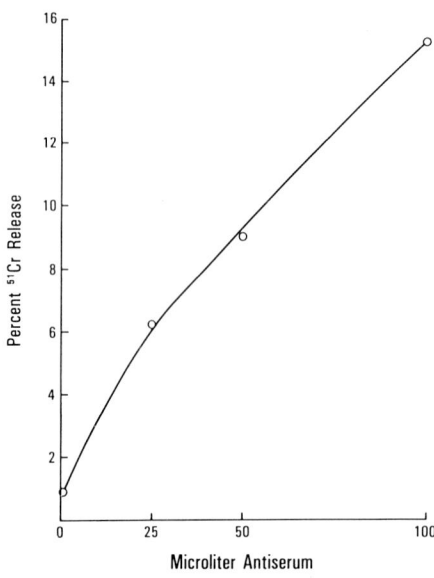

Fig. 5. Enhancement of reactive by of monocytes by the addition of anti-HRF. Monocytes (^{51}Cr-labeled) were heated with various amounts of anti-HRF, washed, and subjected to reactive lysis with human C5b-9. The percent lysis was plotted vs the amount of antibody.

chosen with human large granular lymphocytes (LGL) as effectors and erythrocytes as targets (ZALMAN et al. 1987b). When IgG-sensitized sheep erythrocytes (IgG E_s) were used as targets in a 4 h killing reaction with an equal number of human LGL, the killing was over 30%. With the addition of about 3000 molecules of HRF into the target membrane, the killing was about 12%. The addition of an equal number of molecules of DAF had no effect on the lysis of the IgG E_s. Lysis of the IgG E_s by the isolated perforin protein of human lymphocytes could also be inhibited by cell bound HRF. These observations suggest that HRF can inhibit channel formation not only by the complement proteins C5b-9 but also by the cytolytic protein(s) of cytotoxic lymphocytes.

9 Cell Surface HRF in Cytotoxic Lymphocytes

Stimulation of peripheral blood mononuclear cells (PBMC) with the anti-CD3 monoclonal antibody OKT3 has been shown to result in proliferation of $CD3^+$ lymphocytes, with release of IL-2 and expression of IL-2 receptors (WEISS et al. 1986). Treatment of these cells with OKT3 also induces cytotoxicity in the $CD8^+$ sub-population (JUNG et al. 1986, 1987). OKT3 treatment also induces cell surface HRF by $CD4^+$ and $CD8^+$ T lymphocytes (MARTIN et al. 1988). Only about 10% of the PBMC expressed HRF prior to stimulation; these belonged mainly to the natural killer cell lineage. After 3 days of stimulation, about 90% of the PBMC were HRF-positive. Whereas untreated PBMC were relatively susceptible to lysis by C5b-9 or by isolated perforin, OKT3-stimulated PBMC were largely resistant to both of these agents. This relative resistance could be diminished by the blocking of surface HRF with $F(ab')_2$ anti-HRF, suggesting that resistance was due to the presence of HRF on the lymphocyte surface (MARTIN et al. 1988). Using Sepharose bound anti-HRF, a 65 kDa protein was isolated from stimulated PBMC. This protein was capable of inserting into sheep erythrocytes and rendering them resistant to reactive lysis by C5b-9.

10 HRF from Cytotoxic Granules

If HRF is able to regulate the lytic action of perforin, the cytoplasmic granules of LGL, which contain the cytotoxic perforin, might be expected to contain HRF. Cytoplasmic granules from IL-2-activated lymphocytes were isolated and lysed in high salt. The soluble fraction was passed over an anti-HRF column and the bound protein eluted with high salt. This protein, granule HRF (HRF-G), was the same size as all of the other forms of HRF. HRF-G, when added to chicken erythrocytes, was able to inhibit their reactive lysis by 60%. It was also able to inhibit LGL-mediated cytotoxicity in a 4 h chromium release assay with ^{51}Cr-labeled M21 melanoma cells as targets. The inhibition of cellular cytotoxicity was dose-dependent and reached a maximum of over 90% (ZALMAN et al. 1988).

11 Protection Against Cell-Mediated Cytotoxicity

The results presented above indicate that HRF may be operative in controlling cell-mediated cytotoxicity. This conclusion has been disputed by several investigators. LICHTENHELD et al. (1988) demonstrated that human and mouse perforin can lyse human, mouse, and rabbit erythrocytes, showing an apparent lack of homologous restriction of perforin. Similarly, JIANG et al. (1988), using a wide variety of erythrocyte targets and rat and human lymphokine-activated killer (LAK) cells, noted a lack of species restriction of LAK and perforin-mediated lysis. Two groups, using PNH erythrocytes which are known to be deficient in HRF and normal erythrocytes, concluded that both sets of cells are equally susceptible to cell-mediated lysis (HOLLANDER et al. 1989; KRÄHENBÜHL et al. 1989). HOLLANDER et al. (1989) have also made use of a mutant cell line which has a defect in the biosynthesis of the (GPI) anchor and therefore does not express GPI-anchored proteins. The results show that mutant cells were more easily killed by homologous complement, but that both mutant and wild-type cells were equally susceptible to cell-mediated lysis, suggesting that HRF (or any other GPI-anchored protein) has no effect on cell-mediated lysis.

The ability of HRF to inhibit cellular killing was examined further. HRF-M, derived from human erythrocyte membranes, was inserted into chicken erythrocytes (Ec). These cells were used as targets for reactive lysis with C5b6, C7, and either rat or human serum-EDTA as a source of the complement proteins C8 and C9. As expected, as more M-HRF was inserted into the E_c, lysis by human serum-EDTA was inhibited, whereas lysis by rat serum-EDTA was unaffected (Table 2) (ZALMAN et al. 1991). These same E_c, coated with anti-E_c IgG antibody, were used as targets for an (ADCC) reaction with either rat or human LAK cells. In both cases, HRF (human) was able to inhibit lysis by LAK cells of both species (ZALMAN et al. 1991). Two conclusions can be reached from these results. First, HRF-M can

Table 2. Inhibition of complement- and cell-mediated cytolysis by HRF

HRF-M offered (µg)	Percent inhibition of lysis			
	Human complement	Rat complement	Human LAK cells	Rat LAK cells
0	0	0	0	0
0.15	15	1	56	43
0.30	88	1	81	86

inhibit both human and rat LAK cell killing. Therefore, this inhibition is not species-restricted. Second, HRF can inhibit lysis by human C8 and C9 but not by rat C8 and C9. Inhibition of complement-mediated lysis, in contrast to cellular killing, is species-restricted. HRF, then, is capable of homologous species restriction in the complement system but is a more general inhibitor of LAK cell killing. These results are in agreement with the data presented by several groups, that erythrocytes of several species are equally good targets for LAK cells and purified perforin (LICHTENHELD et al. 1988; JIANG et al. 1988). It has been reported that normal human erythrocytes and PNH-E are approximately equal in their susceptibility to human LGL-mediated LGL killing (KRÄHENBÜHL et al. 1989; HOLLANDER et al. 1989). It is known that PNH cells are missing HRF (ZALMAN et al. 1987a) and therefore might be expected to be more sensitive to ADCC. However, it is possible that the amount of HRF necessary to inhibit LAK cell killing is higher than the 1000 or so molecules of HRF present on a normal erythrocyte. Preliminary evidence indicates that more HRF is needed to inhibit cell-mediated killing than to inhibit complement lysis.

12 Conclusion

HRF is a normal component of erythrocyte and leukocyte membranes. It inhibits lysis caused by homologous complement and therefore protects these blood cells against "innocent bystander lysis" (GÖTZE and MÜLLER-EBERHARD 1970) at sites of complement activation. PNH, a disease in which cell lack HRF, is characterized by an abnormally high amount of hemolysis. The inhibition of cell-mediated cytotoxicity by HRF remains controversial; therefore the complete functional role of HRF is yet to be defined.

Acknowledgement. I would like to thank Tricia Gerrodette for typing the manuscript.

References

Dacie JV, Richardson N (1943) The influence of pH on in vitro haemolysis in nocturnal hemoglobinuria. J Pathol Bacteriol 55: 375

Davé SJ, Sodetz JM (1990) Regulation of the membrane attack complex of complement—evidence that C8γ is not the target of homologous restriction factors. J Immunol 144: 3087–3090

Davies A, Simmons DL, Hale G, Harrison RA, Tighe H, Lachmann PJ, Waldmann H (1989) CD59, an LY-6-like protein expressed in human lymphoid cells, regulates the action of the complement membrane attack complex on homologous cells. J Exp Med 170: 637–654

Davitz MA, Low MG, Nussenzweig V (1986) Release of decay-accelerating factor (DAF) from the cell membrane by phosphotidyl-inositol specific phospholipase C (PIPLC). J Exp Med 163: 1150–1161

Götze O, Müller-Eberhard HJ (1970) Lysis of erythrocytes by complement in the absence of antibody. J Exp Med 132: 898–903

Hänsch GM (1988) The homologous species restriction of the complement attack: structure and function of the C8 binding protein. In: Podack ER (ed) Cytotoxic effector mechanisms. Springer, Berlin Heidelberg New York, pp 109–118 (Current topics in microbiology and immunology, vol 140)

Hänsch GM, Hammer CH, Vanguin P, Shin ML (1981) Homologous species restriction in lysis of erythrocytes by terminal complement proteins. Proc Natl Acad Sci USA 78: 5118–5121

Hänsch GM, Schönermark S, Roelcke D (1987) Paroxysmal nocturnal hemoglobinuria type III—lack of an erythrocyte membrane protein restriction the lysis by C5b-9. J Clin Invest 80: 7–12

Hänsch GM, Wella PF, Nicholson-Weller A (1988) Release of C8 binding protein (C8bp) from the cell membrane by phosphotidyl-inositol-specific phospholipase C. Blood 72: 1089–1092

Holguin MH, Fredrick LR, Bernshaw NJ, Wilcox LA, Parker LJ (1989) Isolation and characterization of a membrane protein from normal human erythrocytes that inhibits reactive lysis of the erythrocytes of paroxysmal nocturnal hemoglobinuria. J Clin Invest 34: 7–17

Hollander H, Shin ML, Rosse WF, Springer TA (1989) Distinct restriction of complement- and cell-mediated lysis. J Immunol 142: 3913–3916

Hu VW, Nicholson-Weller A (1985) Enhanced complement-mediated lysis of type III paroxysmal nocturnal hemoglobinuria erythrocytes involves increased C9 binding and polymerization. Proc Natl Acad Sci USA 82: 5520–5524

Jiang S, Persechini PM, Zychlinsky A, Chau-Ching L, Perussia B, Young JD-E (1988) Resistance of cytolytic lymphocytes to perforin-mediated killing—lack of correlation with complement-associated homologous species restriction. J Exp Med 168: 2207–2219

Jung G, Honsik CJ, Reisfeld RA, Müller-Eberhard HJ (1986) Activation of human peripheral blood mononuclear cells by anti-T3: killing of tumor target cells coated with anti-target-T3 conjugates. Proc Natl Acad Sci USA 83: 4479–4483

Jung G, Martin DE, Müller-Eberhard HJ (1987) Induction of cytotoxicity in human peripheral blood mononuclear cells by monoclonal antibody OKT3. J Immunol 139: 639–644

Krähenbühl OP, Peter HH, Tschopp J (1989) Absence of homologous restriction factor does not affect CTL-mediated cytolysis. Eur J Immunol 19: 217–219

Lachmann PJ, Thompson RA (1970) Reactive lysis: the complement lysis of unsensitized cells II: the characterization of activated reactor as C56 and the participation of C8 and C9. J Exp Med 131: 643–657

Lichtenheld MG, Olsen KP, Lu P, Lowry D, Hameed A, Hengartner H, Podack ER (1988) Structure and function of human perforin. Nature 335: 448–551

Martin DE, Zalman LS, Müller-Eberhard HJ (1988) Induction of expression of cell-surface homologous restriction factor upon anti-CD3 stimulation of human peripheral lymphocytes. Proc Natl Acad Sci USA 85: 213–217

Medof ME, Walter EI, Rutgers JL, Knowles DM, Nussenzweig V (1987) Identification of the complement decay-accelerating factor (DAF) on epithelium and glandular cells and in body fluids. J Exp Med 165: 848–864

Nicholson-Weller A, Burger J, Fearon DT, Weller PF, Austen KF (1982) Isolation of human erythrocyte membrane glycoprotein with decay-accelerating activity for C3-convertases of the complement system. J Immunol 129: 184–189

Nicholson-Weller A, March JP, Rosenfeld SI, Austen KF (1983) Affected erythrocytes of patients with paroxysmal nocturnal hemoglobinuria are deficient in the complement regulatory protein, decay-accelerating factor. Proc Natl Acad Sci USA 80: 5066–5070

Okada N, Harada R, Fujita T, Okada H (1989) A novel membrane glycoprotein capable of inhibiting membrane attack by homologous complement. Int Immunol 1: 205–208

Packman CH, Rosenfeld SI, Jenkins DE, Thiem PA, Leddy JP (1979) Complement lysis of human erythrocytes. Differing susceptibility of two types of paroxysmal nocturnal hemoglobinuria cells to C5b-9. J Clin Invest 64: 428–433

Pangburn MK, Schreiber RD, Müller-Eberhard HJ (1983a) Deficiency of an erythrocyte membrane protein with complement regulatory activity in paroxysmal nocturnal hemoglobinuria. Proc Natl Acad Sci USA 80: 5430–5434

Pangburn MK, Schrieber RD, Trombold JS, Müller-Eberhard HJ (1983b) Paroxysmal nocturnal hemoglobinuria: deficiency in factor H-like functions of the abnormal erythrocytes. J Exp Med 57: 1971

Podack ER, Preissner KT, Müller-Eberhard HJ (1984) Inhibition of C9 polymerization within the SC5b-9 complex of complement by S-protein. Acta Pathol Microbiol Immunol Scand [C] 284: 89–96

Rosenfeld SI, Jenkins DE, Leddy SP (1985) Enhanced reactive lysis of paroxysmal nocturnal hemoglobinuria erythrocytes by C5b-9 does not involve increased C7 binding on cell-bound C3b. J Immunol 134: 506–511

Rosse WF (1973) Variations in the red cells in paroxysmal nocturnal hemoglobinuria. Br J Hematol 24: 327–342

Schönermark S, Rauterberg EW, Shin ML, Löke S, Roelcke D, Hänsch GM (1986) Homologous species restriction in lysis of human erythrocytes: a membrane-derived protein with C8-binding capacity functions as an inhibitor. J Immunol 136: 1772–1776

Schönermark S, Felsinger S, Berger B, Hänsch GM (1988) The C8-binding protein of human erythrocytes: interaction with the components of the complement-attack phase. Immunology 63: 585–590

Shin ML, Hänsch G, Hu VW, Nicholson-Weller A (1986) Membrane factors responsible for homologous species restriction of complement-mediated lysis: evidence for a factor other than DAF operating at the stage of C8 and C9. J Immunol 136: 1777–1782

Sugita Y, Nakano Y, Tomita M (1988) Isolation from human erythrocytes of a new transmembrane protein which inhibits the formation of complement transmembrane channels. J Biochem 104: 633–637

Watts MJ, Dankert JR, Morgan BP (1990) Isolation and characterization of a membrane-attack-complex-inhibiting protein present in human serum and other biological fluids. Biochem J 265: 471–477

Weiss A, Imboden J, Hardy K, Manger B, Terhorst C, Stobo J (1986) The role of the T3/antigen receptor complex in T-cell activation. Annu Rev Immunol 4: 593

Yamamoto K-I (1977) Lytic activity of C5-9 complexes for erythrocytes from the species other than sheep: C9 rather than C8-dependent variation in lytic activity. J Immunol 119: 1482–1485

Zalman LS, Wood LM, Müller-Eberhard HJ (1986a) Isolation of a human erythrocyte membrane protein capable of inhibiting expression of homologous complement transmembrane channels. Proc Natl Acad Sci USA 83: 6975–6979

Zalman LS, Brothers MA, Chui FJ, Müller-Eberhard HJ (1986b) Mechanism of cytotoxicity of human large granular lymphocytes: relationship of the cytotoxic lymphocyte protein to C8 and C9 of human complement. Proc Natl Acad Sci USA 83: 5262–5266

Zalman LS, Wood LM, Frank MM, Müller-Eberhard HJ (1987a) Deficiency of the homologous restriction factor in paroxysmal nocturnal hemoglobinuria. J Exp Med 165: 572–577

Zalman LS, Wood LM, Müller-Eberhard HJ (1987b) Inhibition of antibody-dependent lymphocyte cytotoxicity by homologous restriction factor incorporated into target cell membranes. J Exp Med 166: 947–955

Zalman LS, Brothers MA, Müller-Eberhard HJ (1988) Self-protection of cytotoxic lymphocytes: a soluble form of homologous restriction factor in cytoplasmic granules. Proc Natl Acad Sci USA 85: 4827–4831

Zalman LS, Brothers MA, Müller-Eberhard HJ (1989) Isolation of homologous restriction factor from human urine—immunochemical properties and biologic activities. J Immunol 154: 1943–1947

Zalman LS, Brothers MA, Strauss KL (1991) Inhibition of cytolytic lymphocytes by homologous restriction factor—lack of species restriction. J Immunol 146: 4278–4281

The Effects of Complement Activation on Platelets

D. V. DEVINE

1	Introduction	101
2	Interaction of Platelets with Components of the Classical Pathway	102
2.1	The Platelet C1q Receptor	103
2.2	Physiological Response to C1q Receptor Occupation	103
2.3	Platelet C1 Inhibitor	104
3	Platelet Response to C3 Convertase Complex Proteins and the Products of C3 Cleavage	104
3.1	Platelet Interaction with C3a	104
3.2	Regulation of C3 Activation by the Platelet	105
3.2.1	Decay Accelerating Factor	105
3.2.2	Membrane Cofactor Protein	106
3.2.3	C3 Receptors	107
3.2.4	Endogenous Factor H	107
4	Effects of Membrane Attack Complex Deposition on Platelet Function	108
4.1	Membrane Regulators of MAC	109
4.1.1	Vitronectin Receptor	109
4.1.2	CD59	110
4.1.3	C8 Binding Protein	110
5	Conclusions	111
References		111

1 Introduction

The platelet is a blood element essential to the maintenance of hemostasis. In order to participate in the formation of a hemostatic plug, the platelet must be able to respond to an array of biochemical signals. These signals may arise from other platelets, endothelial cells, or activated proteins of the coagulation and kininogen pathways. The platelet also exhibits physiological responses to activated proteins of another cascade pathway, complement. The interaction of blood platelets with the complement system is the subject of this review.

Platelets normally circulate in an unstimulated state. At this point in their natural history, they have a relatively smooth discoid appearance. Upon contact

Department of Pathology, University of British Columbia, 2211 Wesbrook Mall, Vancouver, BC V6T 2B5, Canada

with any one of a number of physiological agonists, the platelet undergoes a series of biochemical reactions that manifest themselves in morphological and functional changes in the platelet. Initially, the platelet becomes somewhat rounder as a result of cytoskeletal reorganization and acquires the ability to adhere to damaged endothelium. Secondarily, the platelet undergoes a shape change which is associated with the extension of pseudopodia. This morphological change is accompanied by the ability to form aggregates with other platelets. The aggregation phase of platelet activation is accompanied by the release of platelet granule contents into the surrounding medium.

Activated platelets participate in the coagulation process in several ways, the most important of which is that the phospholipid membrane of the platelet serves as the nidus for the formation of the enzyme complex that generates thrombin from prothrombin. Several points of interaction among the complement, fibrinolytic, and coagulation pathways have been reported (reviewed in BLAJCHMAN and OZGE-ANWAR 1986), some of which result in the generation of bioactive fragments of one cascade by enzymes of another. For example, plasmin will cleave both C3 and C5 producing the anaphylotoxic fragments of both substrates. Therefore, in the course of normal and pathologic hemostasis, platelets are exposed to activated complement proteins which may affect their normal function.

In addition to the everyday exposure of circulating platelets to activated complement proteins, settings may arise in which platelets are exposed to supranormal complement activation. The most obvious of these is autoimmune conditions, such as immune thrombocytopenic purpura and systemic lupus erythematosus, in which patients develop autoantibodies directed against or cross-reactive with platelet membrane antigens. The platelets of these patients are targets for both C3 opsonization and membrane attack complex cytolysis. Platelets also are exposed to increased complement activation in sepsis and certain thrombotic platelet disorders including the hemolytic uremic syndrome.

Platelets, like all endogenous cells in contact with complement proteins, can modulate the action of complement. This is generally achieved by platelet membrane proteins that regulate the complement pathway; however, the platelet also has physiological responses to complement activation that may function in regulation of the latter. The platelet appears to have the potential to respond to most steps in the complement activation sequence, from C1q binding to membrane attack complex regulation.

2 Interaction of Platelets with Components of the Classical Pathway

It has been recognized for several decades that immune mechanisms including complement activation influence the responses and survival of platelets. In platelet-rich plasma, the secondary platelet release reaction seen in response to immune complex deposition is complement-dependent (PFUELLER and LUSCHER 1974a), as is the release reaction measured in the presence of zymosan (PFUELLER and LUSCHER 1974b; ZUCKER and GRANT 1974). With the recent expansion of our understanding of the biochemistry of both complement activation and platelet function, we have gained an increased appreciation of the myriad of effects of complement activation on platelets. These are described in detail below, beginning with the classical pathway components.

2.1 The Platelet C1q Receptor

The interaction between platelets and C1q is apparently mediated through a defined platelet membrane C1q receptor (PEERSCHKE and GHEBREHIWET 1987). Unstimulated platelets bear approximately 4000 C1q binding sites and the equilibrium dissociation constant is $10^{-7}\,M$ (PEERSCHKE and GHEBREHIWET 1987). The binding of C1q to platelets is competitively inhibited by soluble collagen. This observation led to the suggestion that the platelet C1q receptor was likely to be the platelet receptor for collagen. The collagen receptor has been identified as the platelet membrane glycoprotein heterodimer Ia/IIa (NIEUWENHUIS et al. 1986; SANTORO 1986). However, recent data suggest that there may be two distinct receptors. PEERSCHKE and GHEBREHIWET (1990) have demonstrated that purified C1q receptor protein binds preferentially to the collagenous portion of C1q but also binds to collagen-coated surfaces. This molecule has different electrophoretic characteristics from the glycoprotein Ia/IIa complex. The binding of purified C1q receptor protein to either collagen or C1q is magnesium-dependent. Intact platelets bind equally well to C1q- or collagen-coated plates. Interestingly, this interaction appears to also require magnesium ions.

2.2 Physiological Response to C1q Receptor Occupation

The presence of C1q causes one of two opposing responses from the platelet. In the presence of monomeric C1q, the platelet fails to respond to collagen stimulation because the collagen/C1q receptor sites are occupied (PEERSCHKE and GHEBREHIWET 1987). Inhibition of collagen-induced platelet aggregation by 50% was induced by as little as 20 µg/ml C1q. The in vivo physiological significance of C1q inhibition of collagen aggregation is unclear since the plasma concentration of monomeric C1q exceeds 100 µg/ml.

The second effect of C1q on platelet function is the stimulation of the platelet aggregation and release reactions by aggregated C1q (CAZENAVE et al. 1976). This interaction is presumably mediated by the collagen-like portion of the C1q molecule. The basis of the requirement for aggregate binding to initiate signal transduction is not understood.

The significance of platelet C1q receptors is not clear. This may be one means of binding immune complexes to the surface of the platelet. However, the binding of C1q may block potential binding sites for C1r and C1s thereby down-regulating complement activation. It is significant to note that, although platelets may bind up to 4000 molecules of C1q per platelet, there is surprisingly little functional C1 activity associated with either fresh or gel-filtered platelets or with platelets stored up to 5 days in autologous plasma (DEVINE et al., unpublished data).

2.3 Platelet C1 Inhibitor

In addition to inhibiting the activity of C1, C1 inhibitor is a major plasma inhibitor of the contact activation phase of blood coagulation. SCHMAIER and colleagues (1985) demonstrated that platelet α granules contain C1 inhibitor, which is released upon stimulation by thrombin or collagen. It is unknown whether this C1 inhibitor can regulate C1 activation at the platelet surface; however, it is able to inhibit the amidolytic activity of purified kallikrein. The release of C1 inhibitor from platelets may play a role in the regulation of both complement activation and contact activation of the coagulation pathway at the site of thrombus formation.

3 Platelet Response to C3 Convertase Complex Proteins and the Products of C3 Cleavage

The activation of complement at the surface of the platelet leads in the first instance to the generation of C3a and C3b. Both of these C3 fragments interact with the platelet as does the product of C3b degradation, C3dg. Platelet responses to these fragments are varied and conflicting reports may reflect particular in vitro experimental conditions. There is a distinct patchiness to information about the interaction of platelets with the proteins of the early complement pathways other than the catabolic products of C3. For instance, there is virtually nothing known about the interactions of platelets with C2, factor B, C4, or their respective catabolites; however, it is known that the exposure of platelets to purified factor D results in the inhibition of platelet aggregation in response to thrombin (DAVIS and KENNY 1979). Factor D appears to be a competitive inhibitor of thrombin; however, the binding affinity for thrombin is greater than for factor D. This may be due to the presence of thrombin receptors of different affinities on human platelets. Exposure of platelets to factor D does not interfere with the ability of the platelet to respond to arachidonic acid or collagen (DAVIS and KENNY 1979).

3.1 Platelet Interaction with C3a

The platelets of both humans and guinea pigs demonstrate detectable changes in responsiveness in the presence of C3a. For an extensive discussion of the structure and function of C3a, the reader is referred to the excellent review by HUGLI (1990). The exposure of guinea pig platelets to C3a induces both aggregation and granule release (ZANKER et al. 1982). If C3a is converted to C3a-des-arg by the removal of the COOH- terminal arginine, the guinea pig platelet fails to respond by either aggregation or release.

The evidence for human platelet responsiveness to C3a is less clear. In experiments examining the agonist effect of C3a on platelet aggregation in a washed platelet system, POLLEY and NACHMAN (1983) reported that, while C3a does not directly induce the aggregation of platelets, it acts synerigstically with the weak platelet agonist ADP. This enhanced aggregation response was seen only in the subpopulation of study subjects who demonstrated a reduced responsiveness to ADP alone. In addition, there was no differential effect caused by the presence or absence of the terminal arginine of C3a. There was no direct effect on human platelet aggregation or release by C3a alone. This line of investigation has recently been expanded by our laboratory. In agreement with the study of POLLEY and NACHMAN (1983), we have demonstrated that there is no direct aggregating effect of C3a on platelets. However, we found that there was no statistically significant effect of C3a on the platelet response to ADP. In fact, we have demonstrated that C3a is able to suppress the ability of the platelet to either aggregate or release in response to strong agonists such as collagen and arachidonic acid (GYONGYOSSY-ISSA and DEVINE 1989). This suppressive effect on strong agonists and the lack of effect on weak agonists occurred both in washed platelet and plasma systems (GYONGYOSSY-ISSA, TAYLOR, and DEVINE, manuscript in preparation). The mechanism of this suppression remains to be identified.

3.2 Regulation of C3 Activation by the Platelet

Several of the membrane proteins reported to regulate the activity of C3 convertases have been found on the surface of platelets. The studies have, to date, been focused on human platelets; however, there is no evidence to suggest that the distribution of membrane regulatory proteins on, for example, murine blood cells would not be similar to that found in humans.

3.2.1 Decay Accelerating Factor

The detailed description of the structure and function of decay accelerating factor DAF can be found elsewhere in this volume (see Chap. 2). The presence of DAF on human platelets was first reported by NICHOLSON-WELLER et al. (1982). Platelet DAF is similar to neutrophil DAF with respect to apparent molecular weight, as determined by SDS-PAGE (KINOSHITA et al. 1985). It is assumed that the function of DAF on platelets is similar to its function on erythrocytes that is, intrinsic protection of the cell from opsonization by the activity of the convertases C4b2a and C3bBb. Data directly addressing the function of DAF on platelets are scarce. DAF inhibition of the alternative pathway C3 convertase, C3bBb, on

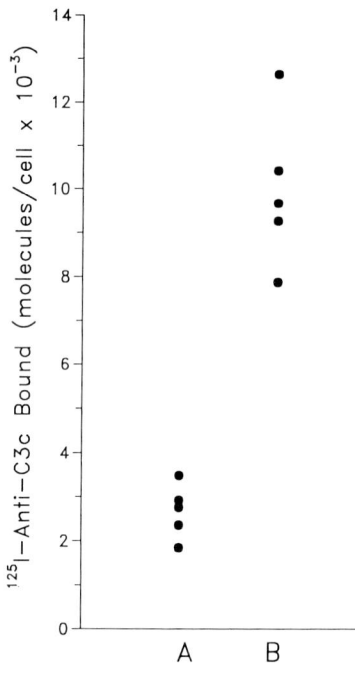

Fig. 1. The regulation of classical pathway convertase C4b2a by platelet DAF. Platelets were isolated by gel filtration and suspended in veronal-buffered saline containing the metabolic inhibitors antimycin A, 2-deoxy-D-glucose, and gluconic acid-*d*-lactone. Platelets were incubated with $F(ab')_2$ fragments of rabbit anti-DAF (B) or normal rabbit serum (A) prior to the addition of cold agglutinin anti-I and C7-deficient human serum. The mixture was incubated first at 4°C then at 37°C. The amount of platelet-bound C3 was determined using ^{125}I-labeled anti-C3c

platelets can be inferred from studies of platelets from patients with paroxysmal nocturnal hemoglobinuria (PNH). The platelets of patients with PNH are deficient in DAF and sometimes demonstrate an increased activity of the platelet-bound C3bBb (DEVINE et al. 1987). Not all DAF-deficient platelets from PNH patients exhibit elevated activity of C3bBb when that enzyme complex is formed using purified complement components and washed platelet suspensions. This may be due to other platelet-specific mechanisms for the regulation of C3bBb discussed below.

The regulation of platelet-bound classical pathway convertase, C4b2a by DAF is demonstrated in Fig. 1. Normal platelets were treated with $F(ab')_2$ fragments of rabbit anti-DAF or nonimmune rabbit IgG prior to incubation in cold agglutinin anti-I and C7-deficient human serum. Inhibition of DAF by anti-DAF resulted in an increase in the amount of C3b detected on the platelet surface. It is reasonable to assume that the conclusions of DAF function studies using erythrocytes can, in the first instance, be extrapolated to platelets.

3.2.2 Membrane Cofactor Protein

Membrane cofactor protein (MCP) is a surface glycoprotein of wide tissue distribution that binds to and modulates the function of C3b. The functional properties of MCP are discussed in detail elsewhere in this volume (see Chap. 4). Although the exact role of platelet MCP in situ remains to be determined, purified platelet MCP functions as a cofactor for the inactivation of C3b, and to a lesser extent C4b, by factor I (SEYA et al. 1986; SEYA and ATKINSON 1989). Interestingly, in the absence of factor I, MCP has a stabilizing effect on the C3 convertases when they are bound to erythrocytes (SEYA and ATKINSON 1989). Similar studies have not been performed for platelets.

3.2.3 C3 Receptors

Platelets can interact with the complement system through receptors specific for C3 or its fragments. Although there is evidence for a C3dg receptor on human platelets (VIK and FEARON 1987; DEVINE and KOVACS 1989), the complement receptors CR1, CR2, CR3 (CD11b/CD18), or p150, 95 (CD11c/CD18) are not present on human platelets. At this time there are no data concerning the physiological effect of exposure to C3dg on platelet function.

Unlike the paucity of information concerning C3 receptors on human platelets, C3 receptors on animal platelets have been fairly well studied. These receptors are believed to function in the clearance of complement-bearing immune complexes, as nonprimate erythrocytes do not express C3b receptors. The platelets of rabbit and guinea pigs bind immune complexes in a complement-dependent manner (TAYLOR et al. 1985), although the physiological effects of this form of complement-platelet interaction remain to be determined. Some clue can be obtained from studies of the interaction of complement-activating liposomes with rat platelets. Infusions of such liposomes into rats cause a transient thrombocytopenia (REINISH et al. 1988). This thrombocytopenia is C3-dependent and is presumably mediated through the platelet C3b receptor. Since the platelet count rapidly returns to normal with no evidence of thrombotic complications, occupation of the C3b receptor presumably is not a stimulus for platelet aggregation or degranulation.

3.2.4 Endogenous Factor H

In addition to DAF, platelets appear to have the ability to regulate the activity of C3bBb by endogenous pools of factor H. The presence of measurable pools of factor H in human platelets was first reported by KENNY and DAVIS (1981). Subsequent studies have demonstrated that, in a system using washed platelets and purified alternative pathway components, the platelet responds to the presence of C3bBb on its surface by releasing its stores of factor H (DEVINE and ROSSE 1987). Platelet factor H is located in the α granules and has an identical electrophoretic mobility to plasma factor H. If the platelet release reaction is blocked by metabolic inhibitors, an increase in the activity of surface-bound C3bBb can be measured. In addition, it is only when the platelet release reaction is blocked that the inhibition of DAF by anti-DAF can be functionally detected (DEVINE and ROSSE 1987).

The potential physiological effects of triggering platelet release by high levels of complement activation are several fold. Activated platelets are thought to play a significant role in the development of pathologic thrombi. It is possible that this type of platelet activation plays a role in the development of thrombocytopenia or disseminated intravascular coagulation in septic patients. The decreased levels of platelet factor H seen in PNH platelets suggest the presence of a population of activated platelets in these patients (DEVINE et al. 1987). This is supported by the concomitant reduction in levels of another platelet granule protein, β-thromboglobulin, as well as the marked predisposition to thrombosis in this group of patients.

Studies of platelet regulation of C3bBb by the release of platelet factor H have yet to be reconciled with studies demonstrating the inhibition of platelet function by C3a. These experimental models differ in several important aspects. Inhibition of platelet-agonist interactions by C3a is affected by the addition of purified C3a to washed platelets or platelet-rich plasma prior to the addition of the agonist. The C3a effect is very likely a charge effect (HUGLI 1990). The release of platelet factor H requires that C3bBb is bound to the platelet surface; it is not triggered by the presence of cobra venom factor-Bb complexes and C3 in the fluid phase. The precise mechanism of release remains to be determined.

4 Effects of Membrane Attack Complex Deposition on Platelet Function

ZIMMERMAN and KOLB (1976) reported the platelet-initiated assembly of membrane attack complex (MAC) in platelet-rich plasma in the absence of known complement activators. The platelet-associated enzyme(s) mediating the direct activation of complement remain to be identified; however, both plasmin and thrombin are candidates. Subsequently, POLLY and NACHMAN (1978, 1979) reported the enhancement of the platelet response to thrombin by simultaneous exposure to complement.

Much of the recent research on the effects of complement activation on platelet function has focused on the MAC. Particularly, the studies of WIEDMER, SIMS, and coworkers have led to a new appreciation of the interactions between the complement and coagulation systems that occur at the platelet surface. Using a system of gel-filtered platelets and purified terminal complement components, these investigators have demonstrated that the deposition of MAC onto the surface of the cell results in the influx of calcium ions into the platelet (WIEDMER and SIMS 1985; WIEDMER et al. 1987). This is associated with the vesiculation of portions of the platelet membrane that are relatively rich in C5b-9 (SIMS and WIEDMER 1986). Vesiculation results in the recovery of the platelet membrane resting potential (SIMS and WIEDMER 1986). In addition, C5b-9 deposition results in the activation of protein kinase C and the release of platelet α granule contents (WIEDMER et al. 1987; ANDO et al. 1988). An increase in thromboxane synthetase has been reported for platelets exposed to C5b-9

(HANSCH et al. 1985; BETZ et al. 1987); however, the platelet release reaction in response to C5b-9 is not blocked by cyclooxygenase inhibitors. Unlike the activation of platelets by other agonists, the release of α granules is not accompanied by a general activation of the platelet. After exposure to C5b-9, the platelet does not express functional fibrinogen receptors; therefore, platelets exposed to C5b-9 show no increase in aggregating ability (ANDO et al. 1988, 1989).

The release of platelet granule contents in response to C5b-9 deposition causes the platelet to be a more effective mediator of coagulation activation (WIEDMER et al. 1986a, b). In the gel-filtered platelet and purified protein system, factor V is released from the α granules and activated to Va by unknown mechanisms. This results in the increased binding of factor Xa to the platelet and the formation of enzymatically active platelet prothrombinase complex. The factor V released from the platelets seems to be disproportionally associated with the platelet vesicles produced in response to C5b-9 (SIMS et al. 1988). The reason for this association is unknown.

Although there are clear effects on platelet function induced by the deposition of MACs onto the platelet surface, it is unclear whether this plays a significant role in immune-mediated platelet destruction or if it is a "bystander" phenomenon. In immune-mediated platelet interactions with complement, it is likely that the platelet would encounter the activated proteins of the early complement pathway before encountering significant amounts of C5b-9. In situations in which complement activation occurs on surfaces other than the platelet membrane, the platelet may respond to the "bystander" deposition of C5b-9. In addition, the studies reported above were conducted in the absence of plasma proteins; therefore, the effect of MAC binding to platelets in a plasma system remains to be determined.

4.1 Membrane Regulators of MAC

Membrane protein regulators of C5b-9 activity have been reported to be present on platelets. They include C8 binding protein, CD59 (MIRL, HRF20), and a platelet receptor for vitronectin. The information concerning the functional activity of these regulatory proteins on platelets is sparse.

4.1.1 Vitronectin Receptor

Platelets bear a well characterized receptor for vitronectin (SUZUKI et al. 1987; GINSBERG et al. 1987). The principal role of this membrane protein is most likely in the binding of platelets to damaged vessels. However, since vitronectin (S protein) is also the primary plasma protein regulator of C5b-9 assembly, it is possible that this receptor can also modulate MAC activity on the platelet surface. Platelets contain an internal pool of vitronectin that is released in response to stimulation by thrombin (PARKER et al. 1989). While the intracellular location of platelet vitronectin has yet to be determined, it is likely that it resides in the platelet α granule. The quantity of vitronectin released from platelets is similar to that reported for the other complement regulatory proteins present in the platelet, namely, factor H (DEVINE and ROSSE 1987) and C1 inhibitor (SCHMAIER et al. 1985). No reports exist concerning the potential modulation of C5b-9 formation by this pool of vitronectin.

4.1.2 CD59

CD59, a membrane regulatory protein of the MAC, is also known as the membrane inhibitor of reactive lysis (MIRL), homologous restriction factor 20 (HRF20), and P18. This protein is present on the surface of normal human platelets. SIMS et al. (1989) investigated the function of platelet CD59 in a plasma-free system. In the presence of blocking antibody to CD59, platelets demonstrated an increased sensitivity to C5b-9. The amount of MAC required to elicit the platelet release response and concomitant vesiculation was approximately ten fold less. The inhibition of CD59 function by antibody had no effect on the lytic capability of C5b-9; that is, there was no increase in cell lysis even though CD59 was blocked. In addition, the ability of C5b-9 to increase platelet prothrombinase activity was affected by platelet exposure to anti-CD59 antibody. These data support the hypothesis that platelet CD59 is able to modulate the platelet responsiveness to C5b-9. The effectiveness of this regulatory protein in platelet-rich plasma remains to be determined. Some inference can be made when considering platelet function in patients with PNH. The platelets of these patients are deficient in CD59 and the pathophysiology of the disease is most frequently related to thrombosis. These two observations are consistent with the conclusions drawn from the in vitro studies described above.

4.1.3 C8 Binding Protein

C8 binding protein has been shown to regulate the activity of MAC on erythrocytes (SCHONERMARK et al. 1986). It has also been reported to be a membrane constituent of normal platelets (BLAAS et al. 1988); however, the electrophoretic mobility of the platelet form of the protein differs from that found on erythrocytes. Reconstitution of C8 binding protein-deficient PNH platelets with purified platelet C8 binding protein reduced the release of serotonin or thromboxane B_2 in response to C5b-9 deposition (BLAAS et al. 1988). It is not known whether C8 binding protein affects the amount of C5b-9 bound to the surface of the platelet, but it does appear to modulate the response of the platelet to activated complement.

Using hemolytic assays, TEDESCO et al. (1986) demonstrated functional C8 associated with the surface of human platelets. This C8 was not part of a C5b-9 complex and treatment of the platelets with thrombin did not induce release of C8 from the platelet. It is possible that the association of C8 with the platelet surface is mediated by C8 binding protein although direct experimental evidence is not available. It has also been reported that collagen-stimulated platelets secrete the terminal proteins of complement, especially C8 and C9 (HOULE et al. 1989); however, there is no evidence to demonstrate whether these proteins are contained in platelet granules or associated with the platelet plasma membrane.

5 Conclusions

Blood platelets are affected in many ways by proteins of the complement pathway. These include the destruction or shortened survival of the platelet, the temporary inhibition of the platelets, normal physiological responses, and the enhancement of coagulation reactions. The effects of complement on platelets may include the generation of the pathogenic effects of platelet activation and destruction including bleeding and thrombosis. While there is an ever increasing body of literature on complement-platelet interactions, there are still many aspects of these interactions that remain to be addressed.

References

Ando B, Wiedmer T, Hamilton KK, Sims PJ (1988) Complement proteins C5b-9 initiate secretion of platelet storage granules without increased binding of fibrinogen or von Willebrand factor to newly expressed cell surface GPIIb-IIIa. J Biol Chem 263: 11907–11914

Ando B, Wiedmer T, Sims PJ (1989) The secretory release reaction initiated by complement proteins C5b-9 occurs without platelet aggregation through GPIIb-IIIa. Blood 73: 462–467

Betz M, Seitz M, Hansch GM (1987) Thromboxane B2 synthesis in human platelets induced by the late complement components C5b-9. Int Arch Allergy Appl Immunol 82: 313–316

Blaas P, Berger B, Weber S, Peter HH, Hansch GM (1988) Paroxysmal nocturnal hemoglobinuria: enhanced stimulation of platelets by terminal complement components is related to the lack of C8bp in the membrane. J Immunol 140: 3045–3051

Blajchman MA, Ozge-Anwar AH (1986) The role of the complement system in hemostasis. Prog Hematol 14: 149–160

Cazenave JP, Assimeh SN, Painter RH, Packham MA, Mustard JF (1976) C1q inhibition of the interaction of collagen with human platelets. J Immunol 116: 162–163

Davis AE, Kenny DM (1979) Properdin factor D: effects on thrombin-induced platelet aggregation. J. Clin Invest 64: 721–728

Devine DV, Kovacs R (1989) Characterization of the C3dg receptor of human platelets. Compl Inflamm 6: 355

Devine DV, Rosse WF (1987) Regulation of the activity of platelet-bound C3 convertase of the alternative pathway of complement by platelet factor H. Proc Natl Acad Sci USA 84: 5873 5877

Devine DV, Siegel RS, Rosse WF (1987) Interactions of the platelets in paroxysmal nocturnal hemoglobinuria with complement. J Clin Invest 79: 131–137

Ginsberg MH, Loftus J, Ryckwaert JJ, Pierschbacher MD, Pytela R, Ruoslahti E, Plow EF (1987) Immunochemical and amino-terminal sequences comparison of two cytoadhesins indicate they contain similar or identical beta subunits and distinct alpha subunits. J Biol Chem 262: 5437–5440

Gyongyossy-Issa MIC, Devine DV (1989) Human platelet responses to ADP agonist are decreased by homologous C3a. Compl Inflamm 6: 340

Hansch GM, Gemsa D, Resch K (1985) Induction of prostanoid synthesis in human platelets by the late complement components C5b-9 and channel forming antibiotic nystatin: inhibition of the reacylation of liberated arachidonic acid. J Immunol 135: 1320–1324

Houle JJ, Leddy JP, Rosenfield SI (1989) Secretion of the terminal complement proteins, C5–C9, by human platelets. Clin Immunol Immunopathol 50: 385–393

Hugli TE (1990) Structure and function of CCa anaphylatoxin. In: Lambris JD (Current topics in microbiology and immunology, Vol 153) (ed) The third component of complement. Chemistry and biology. Springer, Berlin Heidelberg New York, pp 181–208

Kenny DM, Davis AE (1981) Association of alternative complement pathway components with human blood platelets: secretion and localization of factor D and beta-1H globulin. Clin Immunol Immunopathol 21: 351–363

Kinoshita T, Medof ME, Silber R, Nussenzweig V (1985) Distribution of decay-accelerating factor in the peripheral blood of normal individuals and patients with paroxysmal nocturnal hemoglobinuria. J Exp Med 162: 75–92

Nicholson-Weller A, Burge J, Fearon DT, Weller PF, Austen KF (1982) Isolation of a human erythrocyte membrane glycoprotein with decay-accelerating activity for C3 convertases of the complement system. J Immunol 129: 184–189

Nieuwenhuis HK, Sakariassen KS, Houdijk WPM, Nievelstein PFEM, Sixma JJ (1986) Deficiency of platelet membrane glycoprotein Ia associated with a decreased platelet adhesion to subendothelium: a defect in platelet spreading. Blood 68: 692–695

Parker CJ, Stone OL, White VF, Bernshaw NJ (1989) Vitronectin (S protein) is associated with platelets. Br J Haematol 71: 245–252

Peerschke EIB, Ghebrehiwet B (1987) Human blood platelets possess specific binding sites for C1q. J Immunol 138: 1537–1541

Peerschke EIB, Ghebrehiwet B (1990) Platelet C1q receptor interactions with collagen and C1q-coated surfaces. J Immunol 145: 2984–2988

Pfueller SL, Luscher EF (1974a) Studies of the mechanisms of the human platelet release reaction induced by immunologic stimuli I. Complement-dependent and complement-independent reactions. J Immunol 112: 1201–1210

Pfueller SL, Luscher EF (1974b) Studies of the mechanisms of the human platelet release reaction induced by immunologic stimuli. II. The effects of zymosan. J Immunol 112: 1211–1218

Polley MJ, Nachman RL (1978) The human complement system in thrombin-mediated platelet function. J Exp Med 147: 1713–1726

Polley MJ, Nachman RL (1979) Human complement in thrombin-mediated platelet function: Uptake of the C5b-9 complex. J Exp Med 150: 633–645

Polley MJ, Nachman RL (1983) Human platelet activation by C3a and C3a des arg. J Exp Med 158: 603–615

Reinish LW, Bally MB, Loughrey HC, Cullis PR (1988) Interactions of liposomes and platelets. Thromb Haemost 60: 518–523

Santoro SA (1986) Identification of a 160 000 dalton platelet membrane protein that mediates the initial divalent cation-dependent adhesion of platelets to collagen. Cell 46: 913–919

Schmaier AH, Smith PM, Colman RW (1985) Platelet C1 inhibitor: a secreted alpha-granule protein. J Clin Invest 75: 242–250

Schonermark S, Rauterberg EW, Shin ML, Loke S, Roelcke D, Hansch GM (1986) Homologous species restriction in lysis of human erythrocytes: a membrane derived protein with C8-binding capacity functions as an inhibitor. J Immunol 136: 1772–1776

Seya T, Atkinson JP (1989) Functional properties of membrane cofactor protein of complement. Biochem J 264: 581–588

Seya T, Turner J, Atkinson JP (1986) Purification and characterization of a membrane protein (gp45–70) which is a cofactor for cleavage of C3b and C4b. J Exp Med 163: 837–855

Sims PJ, Wiedmer T (1986) Repolarization of the membrane potential of blood platelets after complement damage: evidence for a Ca^{++}-dependent exocytotic elimination of C5b-9 pores. Blood 68: 556–561

Sims PJ, Faioni EM, Wiedmer T, Shattil SJ (1988) Complement proteins C5b-9 cause release of membrane vesicles from the platelet surface that are enriched in the membrane receptor for coagulation factor Va and express prothrombinase activity. J Biol Chem 263: 18205–18212

Sims PJ, Rollins SA, Wiedmer T (1989) Regulatory control of complement on blood platelets. Modulation of platelet procoagulant responses by a membrane inhibitor of the C5b-9 complex. J Biol Chem 264: 19288–19235

Suba EA, Csako G (1976) C1q (C1) receptor on human platelets: inhibition of collagen induced platelet aggregation by C1q (C1) molecules. J Immunol 117: 304–309

Suzuki S, Argraves WS, Arai H, Languino LF, Pierschbacher MD, Ruoslahti E (1987) Amino acid sequence of the vitronectin receptor alpha subunit and comparative expression of adhesion receptor mRNAs. J Biol Chem 262: 14050–14058

Taylor RP, Kujala G, Wilson K, Wright E, Harbin A (1985) In vivo and in vitro studies of the binding of antibody/dsDNA immune complexes to rabbit and guinea pig platelets. J Immunol 134: 2550–2558

Tedesco F, Densen P, Villa MA, Presani G, Roncelli L, Rosso di San Secondo VEM (1986) Functional C8 associated with human platelets. Clin Exp Immunol 66: 472–480

Vik DP, Fearon DT (1987) Cellular distribution of complement receptor type 4 (CR4): expression on human platelets. J Immunol 138: 254–258

Wiedmer T, Sims PJ (1985) Effect of complement proteins C5b-9 on blood platelets: evidence for reversible depolarization of the membrane potential. J Biol Chem 260: 8014–8019

Wiedmer T, Esmon CT, Sims PJ (1986a) Complement proteins C5b-9 stimulate procoagulant activity through platelet prothrombinase. Blood 68: 875–880

Wiedmer T, Esmon CT, Sims PJ (1986b) On the mechanism by which complement protein C5b-9 increases platelet prothrombinase activity. J Biol Chem 261: 14587–14592

Wiedmer T, Ando B, Sims PJ (1987) Complement C5b-9-stimulated platelet secretion is associated with a Ca^{2+}-initiated activation of cellular protein kinases. J Biol Chem 262: 13674–13681

Yu GH, Holers VM, Seya T, Ballard L, Atkinson JP (1986) Identification of a third component of complement-binding glycoprotein of human platelets. J Clin Invest 78: 494–501

Zanker B, Rasokat H, Hadding U, Bitter-Suermann D (1982) C3a induced activation and stimulus specific reversible desensitization of guinea pig platelets. Agents Action 11: 147–157

Zimmermann TS, Kolb WP (1976) Human platelet-initiated formation and uptake of the C5-9 complex of human complement. J Clin Invest 57: 203–211

Zucker MB, Grant RA (1974) Aggregation and release reaction induced in human blood platelets by zymosan. J Immunol 112: 1219–1230

Effects of the Membrane Attack Complex of Complement on Nucleated Cells

B. P. MORGAN

1	Introduction	115
2	Resistance of NC to Lysis by the MAC	116
2.1	Development of the Concept	116
2.2	Lability of the MAC on NC	118
2.3	Removal of the MAC as a Recovery Strategy	118
2.4	Intracellular Signals for MAC Elimination	121
2.5	Membrane Events Involved in Recovery	125
2.6	Influence of MAC Inhibitory Proteins on Recovery	126
3	Nonlethal Effects of the MAC on NC	127
3.1	Background	127
3.2	Release of Reactive Oxygen Metabolites	128
3.3	Eicosanoid Synthesis and Release	128
3.4	Cytokine Release	130
3.5	Other Nonlethal Effects of the MAC	131
3.6	Signaling of Nonlethal Effects of the MAC	131
4	Pathogenic Relevance of Cell Stimulation by the MAC	133
4.1	The MAC and Disease	133
4.2	The MAC in Experimental Diseases	134
4.3	The MAC in Human Diseases	134
	References	135

1 Introduction

In the last quarter of the nineteenth century several workers described the heat-labile lytic action of serum on bacteria and erythrocytes, and it was these observations which led to the discovery of the complement system (NUTTAL 1888; BORDET 1898; EHRLICH and MORGENROTH 1899). These lytic activities were subsequently shown to be mediated by the final stage in the complement system, the membrane attack complex (MAC). Given this history it is therefore perhaps not surprising that, until very recently, the MAC was considered by the majority of immunologists to be a lytic entity, the sole role of which was kill target cells. The concept that the MAC might cause more subtle (and often more pathologically relevant) changes in target cells has only recently gained widespread

---

Department of Medical Biochemistry, University of Wales College of Medicine, Cardiff CF4 4XN, Wales, UK

acceptance. The concept is of particular relevance when the targets are nucleated and metabolically active, although important nonlethal changes may also be induced in non-nucleated cells (see Chap. 7).

My aim in this chapter is to summarize the current understanding of the recovery mechanisms and nonlethal effects initiated in metabolically active nucleated cells (NC) by the MAC. I will detail the types of changes observed in cells in vitro and attempt to unravel the complex intracellular signaling pathways by which the MAC brings about these changes. The structure and mechanism of assembly of the MAC and the role of membrane inhibitory proteins in controlling its lytic effects on cells are detailed elsewhere in this volume and will not be reiterated here. In the final section of the chapter I will relate these nonlethal changes, most of which have been observed only in vitro, to disease situations in which cell stimulation by the MAC may be of pathogenic importance.

2 Resistance of NC to Lysis by the MAC

2.1 Development of the Concept

The aged erythrocyte has been an extremely valuable tool in the study of the complement system. The sensitivity of this metabolically inert target to lysis by complement has made it possible to identify the components involved and their interactions with each other and with the membrane (MAYER 1972; MAYER et al. 1983). However, even in the earliest studies of complement lysis, it was recognized that NC were far more difficult to kill with complement than were erythrocytes, yet the mechanisms underlying this resistance were little studied until the late 1950s. At that time two distinct schools of thought emerged: those who ascribed resistance to an inefficiency of complement activation on the cell surface and those who postulated the existence of active recovery processes within NC, allowing them to escape lysis. Inefficient activation of complement on the cell surface was postulated to be the result of low antigen density, and thus little C1 fixation, or to a reduced efficiency of binding of the late components (MOLLER and MOLLER 1962; LINSCOTT 1970). This hypothesis was not supported by subsequent work, which examined the variation in susceptibility to complement lysis between different cell lines or between cells at different stages of the cell cycle. These studies revealed no consistent relationship between antigen density (CIKES 1970; FERRONE et al. 1973; PELLEGRINO et al. 1974) or the amount of early or late complement components (OHANIAN and BORSOS 1975; COOPER et al. 1974) bound on the membrane and the degree of lysis but did indicate an association between cell metabolic activity and survival. An elegant series of studies examining complement attack on a nucleated tumor cell line by GOLDBERG, GREEN, and colleagues provided substantial evidence for the existence of active resistance mechanisms, at least in the cell line used as target (GOLDBERG and GREEN 1959; GREEN et al. 1959; GREEN and GOLDBERG 1960). These studies showed that complement attack caused cell swelling and release of intracellular amino acids, nucleic acids, and ions, yet larger molecules were retained, demonstrating that membrane integrity was not completely lost. From these results it was concluded that MAC-mediated complement injury could occur in NC in the absence of lysis.

By the early 1970s it was therefore clear that NC could survive limited complement membrane attack and that survival was, at least in part, dependent on metabolic responses within the cell. The nature of the metabolic changes which permitted cell survival were, however, completely unknown. In an attempt to elucidate the important metabolic changes, OHANIAN and coworkers attacked NC with complement in the presence of a wide range of metabolic inhibitors. They found that treatment of guinea pig hepatoma cells with antitumor drugs such as puromycin, adriamycin and actinomycin D, either alone or in combination, increased cell lysis on subsequent exposure to complement, whereas treatment with anabolic hormones such as insulin and hydrocortisone diminished lytic susceptibility (SEGERLING et al. 1975a; SCHLAGER et al. 1976). Inhibition of mitochondrial function and hence cellular energy production with the respiratory chain poisons sodium cyanide or sodium azide caused only limited enhancement of complement-mediated cytolysis (SEGERLING et al. 1975b).

All of the agents which significantly affected cytolysis were capable of influencing lipid metabolism and this, together with the kinetics of the observed changes, led to the conclusion that alterations in cellular lipid content and turnover were central events in the modulation of lytic susceptibility (OHANIAN and SCHLAGER 1981). Although the mechanisms by which changes in lipid content or metabolism altered lytic susceptibility were not clear from these studies, it was suggested that membrane repair mechanisms existed which were influenced by these changes. Evidence in favor of this possibility was provided by studies in which the membrane lipid compositions of liposomes, erythrocytes, or NC were artificially altered (SHIN et al. 1978; YOO et al. 1980; OHANIAN et al. 1979; 1982). Alteration of membrane lipid content modulated the lytic susceptibility of the target, but no consensus emerged regarding the specific lipids confering complement resistance or the mechanisms by which resistance was brought about.

A second group of agents which were shown to cause significant enhancement of complement lysis of NC were enzymes, including neuraminidase, trypsin, pepsin, and pronase (BOYLE et al. 1978). Neuraminidase strips sialic acid from the membrane, making the cell a better complement activator. It is likely that the proteolytic enzymes strip nonspecific or specific (cell surface inhibitors) protein protective factors from the cell surface, although definitive proof of this is lacking.

2.2 Lability of the MAC on NC

The aged sheep erythrocyte, the classical target for assessing complement lytic activity, is little more than a membrane bag containing proteinaceous fluid. Formation of transmembrane pores in these targets leads to an osmotically driven influx of water, causing the cell to swell and eventually burst. The lytic process is highly efficient, only a single functional lesion being required to bring about cell lysis (single-hit) (MAYER 1961). Although lysis is efficient, lesion formation is very inefficient; there are many hundreds of MACs present on the cell at the minimum lytic dose but most are ineffectual. Functional pores formed by the MAC on erythrocyte ghosts or on liposomes have been shown to be highly stable, remaining in the membrane for several days after formation (RAMM and MAYER 1980; RAMM et al. 1983a). In marked contrast, MAC pores on NC do not inevitably cause cell lysis and are transient. Pores which allow an efflux of small markers, such as rubidium ions, or an influx of calcium ions have been formed on NC without lysis ensuing, implying that the cell can tolerate some degree of membrane leakiness (BOYLE et al. 1976a; HALLETT et al. 1981). Further, it has been shown that NC can be "rescued" from a potentially lytic dose of the MAC by incubating with cAMP (KALINER and AUSTEN 1974; BOYLE et al. 1976b) suggesting that pores can be closed or removed from the membrane. Lability of the MAC-induced pore was first demonstrated in isolated nerve and muscle cells by patch-clamping studies during complement attack (STEPHENS and HENKART 1979; JACKSON et al. 1981). Individual pores, detected by membrane depolarization, appeared sporadically in the membrane and, after a short but variable interval, disappeared. Further evidence of pore lability was provided by kinetic analysis of cell killing using the Molt 4 cell line (KOSKI et al. 1983). Cytolysis was shown to obey multi-hit kinetics, indicating that many effective lesions were required on the membrane in order to bring about lysis. The transience of lesions in NC was also demonstrated in the tumor cell line U937, in which MAC pores, assessed by measuring the rate of release of entrapped rubidium ions, had a functional lifetime of only a few minutes (RAMM et al. 1983b). Inhibition of ion pumps by puromycin had little effect on the lifespan of pores, implying that, at least in this cell line, increased removal of ions and water from the cell was not a major factor in resisting lysis (RAMM et al. 1984). At this time it was unclear whether MAC pores were being blocked in some way or whether the MAC itself was actually removed from the cell surface. Evidence indicative of physical removal of MACs has subsequently been obtained from a wide range of NC.

2.3 Removal of the MAC as a Recovery Strategy

NC therefore appear to possess escape mechanisms, not present on aged erythrocytes, which allow them to withstand limited complement membrane attack. Functional studies have demonstrated that pores are transient but provide few clues to the mechanisms involved. Disappearance of functional

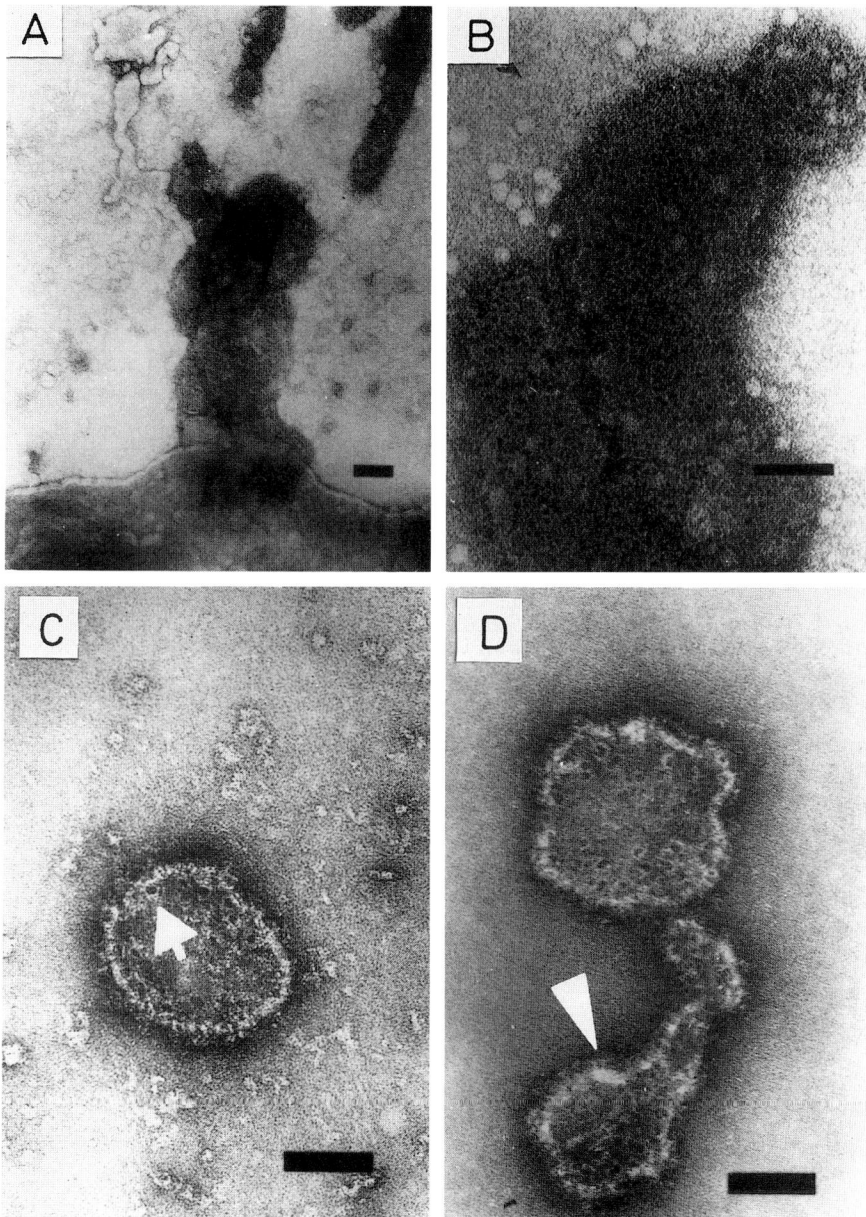

Fig. 1A–D. Vesiculation of neutrophils during membrane attack. **A** Transmission electron micrograph of a neutrophil fixed 2 min after attack with nonlethal amounts of the MAC. The cell surface is devoid of MACs except in the area of the villous projection. **B** Detail of **A** showing densely packed ring lesions on villous projection. **C, D** Vesicles shed from neutrophils during nonlethal complement membrane attack. The surfaces of the vesicles are covered with ring lesions. *Bar* represents 100 nm in each micrograph

pores could be due to inactivation of the MAC, the effete complex remaining on the cell surface, or to physical removal of the MAC from the membrane. Unusual morphological changes in NC during complement attack, including the appearance of multiple membrane protrusions (OHANIAN et al. 1977) and the shedding of vesicles from the cell surface (GOLDBERG and GREEN 1959; RICHARDSON and LUZIO 1980; PODACK and MULLER-EBERHARD 1981), had been noted in diverse cells. An association of these surface changes with recovery from complement membrane attack was first demonstrated in the neutrophil. During nonlethal attack, MACs (measured by binding of specific antibodies) were rapidly removed from the surface of human neutrophils, the half-life of complexes on the membrane being only about 3 min at 37 °C (MORGAN et al. 1984; CAMPBELL and MORGAN 1985). Membrane vesicles shed from the cell surface during attack stained strongly for C9 and so presumably contained MACs. In more detailed studies of the fate of the MAC on neutrophils during nonlethal attack it was shown that an average of 25 000 MACs per cell could be formed on human neutrophils without causing significant lysis (MORGAN et al. 1987). Most of these complexes (65%) were rapidly shed on membrane vesicles which constituted just 2% of the cell surface. The remainder of the cell-bound MACs were internalized and degraded within the cell. Shed vesicles isolated from the supernatant were densely covered with MACs, whereas no MACs were detectable on the cell membrane within 5 min of attack (Fig. 1).

Physical removal of MACs by vesiculation (ectocytosis) and/or endocytosis has now been demonstrated in a large number of cell types (Table 1). In some NC, for example the Ehrlich murine carcinoma cell line, the principal route of elimination appeared to be via endocytosis, MACs being removed at rates comparable to those achieved by ectocytosis in the neutrophil (CARNEY et al. 1985). C5b-8 sites on neutrophils were relatively stable, with a half-life for removal

Table 1. Routes of MAC elimination

Cell	Route	Ca^{2+} dependent	Reference
Neutrophils (rat and human)	Ecto > Endo	√	CAMPBELL and MORGAN 1985; MORGAN et al. 1987
EAT cells	Endo	√	CARNEY et al. 1985; 1986
U937 cells	Ecto	√	MORGAN et al. 1986a
Glomerular epithelial cells (rat)	Ecto	?	CAMUSSI et al. 1987
	Trans	?	KERJASCHKI et al. 1989
Oligodendrocytes (rat)	Ecto	√	SCOLDING et al. 1989b
Rheumatoid synovial cells (human)	Ecto	√	Unpublished
K562 cells	Ecto	√	Unpublished
Platelets	Ecto	√	SIMS and WEIDMER 1986

Ecto, ectocytosis; Endo, endocytosis; Trans, transcytosis

of over 60 min at 37°C. On Ehrlich cells, however, C5b-8 sites were rapidly removed, albeit at a slower rate than C5b-9 sites (half-life about 20 min). Even C5b-7 sites, which presumably caused little or no membrane perturbation, were efficiently cleared from the surface of these cells, suggesting that the events signaling lesion removal were different in the two cell types. However, the Ehrlich cell studies were performed using a heterologous system—murine cells and human complement—and it is possible that this species incompatibility was responsible for the differences observed.

Like the neutrophil, the human histiocytic cell line U937 removed homologous MACs efficiently, primarily via ectocytosis, but C5b-8 sites were stable and, if present on the membrane in sufficient numbers, caused cell killing. Addition of trace amounts of C9 to U937 cells bearing C5b-8 complexes markedly enhanced removal and consequently decreased killing of these cells (MORGAN et al. 1986a). These results suggest that the C5b-8 lesion is an insufficient stimulus for recovery in cells attacked by homologous complement, but may be sufficient in heterologous systems because of the lack of other inhibitory factors. Alternatively, it is possible that the C5b-8 complex is efficiently removed by endocytosis in those cells utilizing this recovery mechanism, but removal by ectocytosis requires the formation of the complete MAC. Rat renal glomerular epithelial cells also remove MACs by vesiculation when attacked in vitro (CAMUSSI et al. 1987), and a transcytotic route of MAC elimination by these cells in vivo has recently been described (KERJASCHKI et al. 1989). Using immunoelectron microscopy the fate of MACs formed in the kidney in experimental glomerulonephritis was monitored. It was shown that MACs were endocytosed at the abluminal surface of the glomerular epithelial cell and then transported across the cell to be exocytosed into the urinary space. MACs were present in the urine of animals with experimental glomerulonephritis, suggesting that transcytosis was a major route of MAC elimination from the kidney in this disease (KERJASCHKI et al. 1989). Preliminary studies in our laboratory have detected MACs in the urine of patients with glomerulonephritis, indicating that similar processes may be occurring in human renal disease (unpublished observations).

Specific mechanisms therefore exist in NC enabling them to recover from nonlethal complement membrane attack by removal of the MAC from the cell membrane. The processes and intracellular signals responsible for MAC removal are now the focus of much research interest.

2.4 Intracellular Signals for MAC Elimination

The demonstration by CAMPBELL and coworkers that the first intracellular event detectable upon formation of the MAC on a cell was a rapid increase in intracellular free calcium concentration ($[Ca^{2+}]_i$) provided an early clue to the intracellular signals responsible for mediating recovery (CAMPBELL et al. 1979; CAMPBELL and LUZIO 1981). An increase in $[Ca^{2+}]_i$ occurred within seconds of adding C9 to cells bearing C5b-8 sites and preceded cell lysis by several minutes

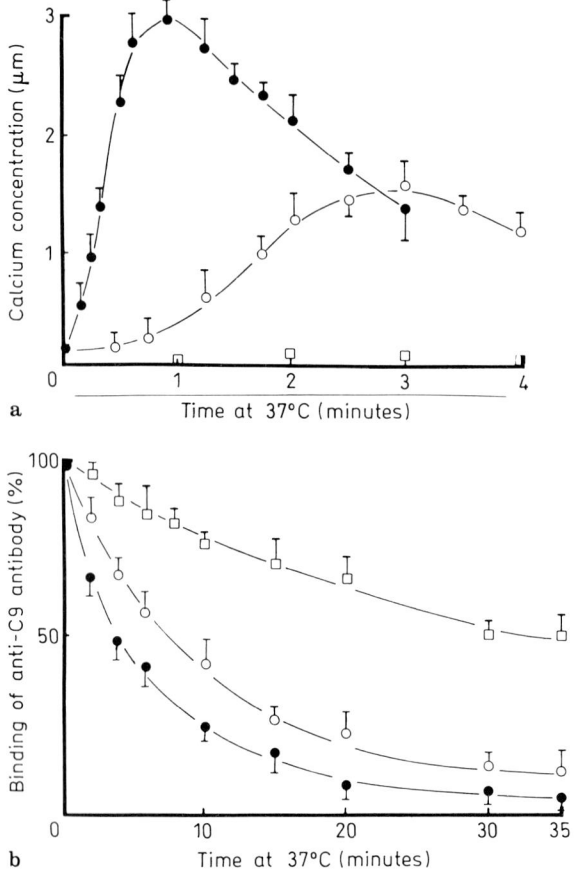

Fig. 2a, b. Calcium-dependence of MAC elimination of neutrophils. Neutrophils containing the Ca^{2+}-activated photoprotein obelin were subjected to membrane attack in the presence of extracellular calcium (1.3 mM, *closed circles*; end point cell death, <5%), in the absence of extracellular calcium (*open circles*; end point cell death, 14%) or after chelation of intracellular calcium (1 mM EGTA, *open squares*; end point cell death, 25%). **a** Intracellular free calcium concentration was calculated from the obelin luminescence at each time point under each set of conditions. **b** Removal of MACs from the cell surface was assessed by measuring the binding of a radiolabeled anti-C9 antibody to the cells at each time point under each set of conditions. (From MORGAN and CAMPBELL 1985)

(CAMPBELL et al. 1981). It was suggested that the rapid rise in the intracellular levels of this important regulator could, under conditions of mild attack, signal MAC removal from the cell surface (CAMPBELL and LUZIO 1981). In the neutrophil the MAC induced transient changes in $[Ca^{2+}]_i$ even in the absence of subsequent lysis, the concentration rising from a resting level of about 0.2 µM to a peak of up to 5 µM within 1 min of MAC formation and the falling rapidly back towards the resting level (MORGAN and CAMPBELL 1985). When extracellular calcium was removed by chelation with EDTA the MAC-induced increase in $[Ca^{2+}]_i$ was delayed and inhibited but not completely lost. Recovery by MAC removal was concomitantly slowed and cell death was enhanced. Chelation of intracellular calcium completely inhibited MAC removal and further enhanced lysis, firmly implicating calcium as a mediator of recovery and demonstrating that the increase in $[Ca^{2+}]_i$ was due not only to influx of calcium via the pore but also to release from intracellular stores (Fig. 2). Calcium is a potentially toxic ion and above a certain threshold level, which probably varies between cell types, the toxic effects of raised $[Ca^{2+}]_i$ will exceed the protective effects and cell lysis will then ensue (MORGAN et al. 1986b). A central role of calcium in signaling MAC removal and cell recovery has subsequently been demonstrated in many NC, including Ehrlich cells, oligodendrocytes, synovial cells, and glomerular epithelial cells and also in platelets (for references see Tabel 1). In Ehrlich cells, the higher the peak $[Ca^{2+}]_i$ reached during nonlethal attack the faster intracellular levels subsequently fell, suggesting an enhancement of recovery and further implicating calcium as the major intracellular signal (CARNEY et al. 1986). Single cell analysis in Ehrlich cells loaded with the calcium binding fluorophore fura-2 during nonlethal attack with serum has demonstrated that the MAC initiates oscillations of $[Ca^{2+}]_i$ (CARNEY et al. 1990), suggesting that feedback regulation of $[Ca^{2+}]_i$ was occurring. We have observed similar oscillations in oligodendrocytes and in neutrophils during nonlethal attack even when new site formation on the cells was prevented by adding C8 and C9 to washed cells bearing a fixed number of C5b-7 sites.

From the above account it is clear that calcium is an important signal for recovery processes in a variety of NC attacked by the MAC. The mechanisms by which an increase in $[Ca^{2+}]_i$ brings about MAC removal are far less clear. Most of the intracellular consequences of raised $[Ca^{2+}]_i$ are mediated via calmodulin and involve the modulation of the activity of protein kinases which phosphorylate, and thus regulate, key enzymes in the cell (CAMPBELL 1982; COX 1988; COLBRAN et al. 1989). MAC-induced protein kinase activation, mediated at least in part by increased $[Ca^{2+}]_i$, has been directly demonstrated in several types of NC (FISCHELSON et al. 1989; CYBULSKY et al. 1990; CARNEY et al. 1990). Inhibition of protein kinases or of calmodulin inhibited MAC removal and consequently increased lysis in these cells, suggesting that recovery was mediated at least in part via these secondary effectors.

An involvement of other intracellular regulators in the mediation of recovery processes has also been suggested. Several studies have implicated cAMP: NC bearing potentially lytic amounts of the MAC were "rescued" from subsequent lysis by incubation with cAMP (KALINER and AUSTEN 1974; BOYLE et al. 1976b); the lytic susceptibility of a guinea pig cell line inversely correlated with the intracellular cAMP concentration (LO and BOYLE 1979); and pharmacological agents which modulated intracellular cAMP levels inhibited MAC removal and enhanced cell lysis (ROBERTS et al. 1985; FISHELSON et al. 1989). The actions of cAMP are mediated via the modulation of specific protein kinases and include the phosphorylation of many cellular proteins, these phosphorylation events

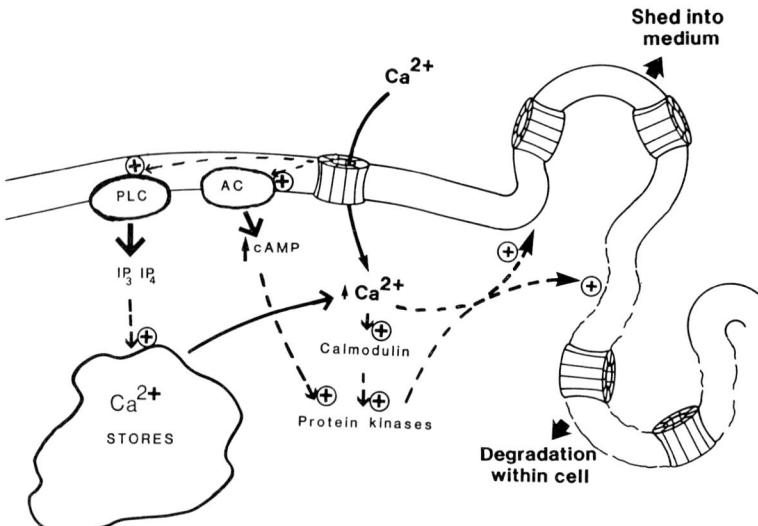

Fig. 3. Signaling of MAC elimination. The intracellular signals implicated in MAC elimination from NC membranes are shown here. Increased intracellular free calcium concentration, caused by influx of calcium through the MAC pore, is the primary signal for MAC removal by ectocytosis or endocytosis. Calcium may have a direct membrane destabilizing effect on the membrane or may act via calmodulin to activate cytosolic and membrane enzymes. The MAC may also directly influence membrane signal generators such as phospholipase C (*PLC*) and adenylate cyclase (*AC*), thus causing increased intracellular levels of mediators including inositol lipids and cAMP which may also signal MAC removal

may be important in signaling recovery (see below). An MAC-induced increase in cAMP has recently been demonstrated in Ehrlich cells (CARNEY et al. 1990), but the mechanism(s) by which the MAC brings about an increase in intracellular cAMP are unclear. Ca^{2+}–calmodulin complexes can modulate the enzymes responsible for cAMP production and breakdown (COX 1988). Thus, increased $[Ca^{2+}]_i$ may bring about a secondary increase in cAMP, thereby enhancing recovery.

A second possibility is that the MAC directly influences membrane signaling systems. In support of this hypothesis is the observation that the MAC causes an increase in $[Ca^{2+}]_i$ in neutrophils and in glomerular epithelial cells even in the absence of extracellular calcium (MORGAN and CAMPBELL 1985; CYBULSKY et al. 1990), presumably due to a signaled release of calcium from stores. One candidate is the G protein system, which couples diverse cell surface receptors with multiple intracellular effectors, including cyclic nucleotides and inositol phospholipids (CASEY and GILMAN 1988; TAYLOR 1990). MAC interaction with G proteins, directly or via a receptor, could thus cause increased cytosolic concentrations of cAMP and calcium and influence other membrane components. It has recently been shown that inhibitors of G proteins diminish the cAMP response and enhance lysis of NC attacked by the MAC (CARNEY et al. 1990;

DANIELS et al. 1990b). However, direct interaction of the MAC with membrane receptor molecules or with G proteins has yet to be demonstrated.

The above account demonstrates that the signaling of recovery in NC is complex, involving multiple intracellular signals including calcium, cyclic nucleotides, and inositol phosphates. Calcium influx via the MAC pore is of central importance, but the MAC may also directly influence membrane signaling systems. NC subjected to nonlethal attack by the MAC release numerous active molecules, including eicosanoids and cytokines (see below), which may also contribute to recovery processes (Fig. 3).

2.5 Membrane Events Involved in Recovery

Multiple intracellular mediators thus act to signal removal of MACs from the membrane. However, the physical processes bringing about ectocytosis or endocytosis of MACs are still unclear. Endocytic elimination of MACs or incomplete complexes from the membrane in Ehrlich cells involves accumulation of complexes in clathrin-coated pits followed by internalization and subsequent degradation in lysosomes (CARNEY et al. 1986). Elimination of MACs by transcytosis in glomerular epithelial cells also appears to utilize clathrin-coated pits and intracellular vesicle transport systems (KERJASCHKI et al. 1989).

MAC removal by ectocytosis is a two-stage process involving their accumulation in densely packed "patches" on the membrane and the subsequent shedding of these MAC-rich areas. The mechanisms causing patching of MACs prior to shedding are unclear, there being no evidence of association of complexes with specific cell surface structures. Aggregation of MACs might occur by adherence of complexes as they meet during their random migrations in the cell membrane, but the absence of patching in aged erythrocytes or liposomes makes this explanation unlikely. Cytoskeletal components, involved in patching and capping of diverse membrane receptors, could cause aggregation of MACs, but inhibitors of microtubule and microfilament activity did not inhibit patching and ectocytosis in neutrophils (MORGAN and DANKERT, unpublished observations).

The shedding stage is no better understood. Large aggregates of MACs might cause vesiculation simply by disturbing the integrity of the surrounding bilayer (MORGAN et al. 1987). However, the abundant evidence described above, demonstrating an involvement of intracellular signaling in recovery, makes this attractive hypothesis untenable. Recent studies of the vesicles shed by neutrophils during MAC attack have shown that sorting of membrane proteins and lipids occurs (STEIN and LUZIO 1991). The shed vesicles contained much higher proportions of cholesterol and diacylglycerol than did the neutrophil plasma membrane. Both of these lipids have been shown to have a destabilizing effect on membranes, diacylglycerol causing increased membrane curvature and membrane fusion when present in high concentrations (ALLAN and MICHELL 1979) and cholesterol having a condensing effect on membrane phospholipids, again favoring increased membrane curvature (LEVINE 1972). Interactions of $[Ca^{2+}]_i$ with cholesterol in the membrane have also been reported to bring about destabilization of model lipid membranes (CHEETHAM et al. 1990). A localized area of high cholesterol content around patches of MACs, together with the MAC-induced increase in $[Ca^{2+}]_i$, could therefore represent a potent stimulus to vesiculation. Localized or generalized increases in $[Ca^{2+}]_i$ can also cause activation of phospholipases (CYBULSKY et al. 1989), which may further contribute to membrane destabilization around the lesion.

Calcium and cyclic nucleotides may also directly influence the MAC. C9, the major component of the MAC, has recently been shown to be a calcium binding protein (THIELENS et al. 1988) and to be susceptible to phosphorylation in vitro (FISCHELSON et al. 1989). Phosphorylation of or calcium binding to C9, brought about by increased intracellular cyclic nucleotide or $[Ca^{2+}]_i$, might therefore enhance MAC elimination by increasing the tendency of MACs to aggregate in the membrane or by increasing MAC-mediated membrane perturbation.

Examination of the metabolic status of target cells after nonlethal membrane attack has revealed depleted energy stores and functional impairment, implying a considerable energy cost in resisting lysis and restoring cellular equilibria (MORGAN 1988; SCOLDING et al. 1989a). These metabolically depleted cells were more susceptible to lysis by a second attack but, in the presence of suitable substrates, energy stores and cell resistance were rapidly restored.

2.6 Influence of MAC Inhibitory Proteins on Recovery

The membrane-associated MAC inhibitory proteins, described by ZALMAN and by HOLGUIN and PARKER elsewhere in this volume, have been shown to incorporate into the forming MAC on erythrocytes (HANSCH 1988; MERI et al. 1990). These proteins have been identified on a variety of NC and recent evidence would suggest that they serve the same inhibitory function as the erythrocyte protein (NOSE et al. 1990; ROONEY and MORGAN 1990a; ROONEY et al. 1991). These findings raise the possibility that the bulk of MACs formed by homologous complement on NC are inactivated by inhibitory proteins prior to their removal from the cell surface. We have recently obtained preliminary evidence that the 20 kDa inhibitory protein CD59 antigen is selectively removed from the cell surface in association with the MAC during recovery in the K562 erythroleukemia cell line (MORGAN and DANIELS, unpublished observations). Surface expression of endogenous CD59 antigen, measured on the fluorescence activated cell sorter, fell by 25% during nonlethal attack, and reincorporated radiolabeled CD59 antigen was shed in an MAC-rich vesicle fraction (Fig. 4). Removal of MACs from the membrane may therefore merely be a garbage disposal system, clearing large numbers of inactivated MACs, perhaps together with the occasional active complex which has escaped the attentions of the inhibitory proteins.

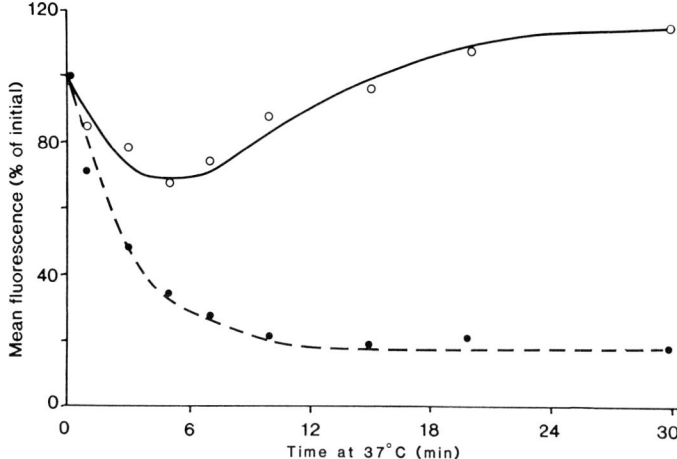

Fig. 4. Removal of CD59 Antigen during nonlethal membrane attack. K562 cells, a CD59-positive human erythroleukemia cell line, were subjected to nonlethal membrane attack by first forming C5b-7 sites and then adding C8 and C9 on ice. The cells were warmed to 37°C and portions were removed at intervals, placed on ice, and fixed in 1% paraformaldehyde. Fixed cells at each time point were then stained with a monoclonal anti-CD59 antibody (*open circles*) or with a monoclonal anti-MAC antibody (*closed circles*). After washing and incubating with fluorescent second antibody the cells were analyzed in a fluorescence activated cell sorter. Mean CD59 and MAC fluorescence was measured at each time point.

3 Nonlethal Effects of the MAC on NC

3.1 Background

Due to the array of protective and recovery mechanisms (described above and elsewhere in this volume), NC are highly resistant to lysis by homologous C5b-9. Nevertheless, complement membrane attack is not without consequence. Stimulatory effects of complement on NC in vitro were noted over 20 years ago and included release of lysosomal enzymes by cartilage cells (LACHMANN et al. 1969) and stimulation of prostaglandin synthesis and bone resorption in fetal long bone cultures (RAISZ et al. 1974). These latter effects were subsequently shown to require formation of the MAC. C6-deficient serum did not cause prostaglandin synthesis or bone resorption; thus, this was the first report of a stimulatory effect of the MAC (SANDBERG et al. 1977). More recently, with the realization that NC usually survive attack by homologous MAC, a great deal of research interest has arisen regarding the nonlethal effects of the complex. Many types of human and nonhuman NC have been chosen as targets, attacked with homologous or heterologous complement, and a host of parameters subsequently measured. From these studies a complex and often contradictory literature has arisen, making it difficult to extract a logical summary. In particular, the use of heterologous systems, the constituents of which often resemble a biochemical zoo, has greatly complicated interpretation. Cells in vivo are exposed to homologous complement and it is the nonlethal effects of homologous attack which are of most relevance to pathogenesis. As a general rule the consequences of nonlethal attack are predominantly proinflammatory. I will here describe some of these consequences and the cell types in which they have been observed in vitro. I will then attempt to distill a concise schema for nonlethal cell stimulation and discuss the likely significance of these nonlethal effects to disease in vivo.

3.2 Release of Reactive Oxygen Metabolites

Reactive oxygen metabolites (ROM), a term which encompasses the many reactive species of oxygen including O_2^-, ·OH·, and H_2O_2, are highly toxic molecules causing a wide range of deleterious effects in cells (HALLIWELL and GUTTERIDGE 1984; SLATER 1984). Phagocytic cells have the capacity to make ROM and use them in the process of intracellular killing of microorganisms. These cells also release ROM which contribute to tissue damage and inflammation at the site of release. Attack by complement has been shown to stimulate release of ROM from phagocytic and, occasionally, nonphagocytic NC. This was first demonstrated using rat and human neutrophils exposed to antibody and human complement (HALLETT et al. 1981) and was subsequently shown to occur in the absence of killing and to require formation of the complete MAC (CAMPBELL and MORGAN 1985; ROBERTS et al. 1985). Release of ROM by nonlethal complement membrane attack has subsequently been observed from several types of NC (Table 2). In all these targets ROM release required assembly of the complete MAC and the presence of extracellular calcium. ROM release did not occur in the absence of C9. Furthermore, chelation of extracellular calcium after the formation of C5b-7 sites on the cell completely inhibited the release of these metabolites on subsequent addition of C8 and C9, suggesting that the rapid MAC-induced increase in $[Ca^{2+}]_i$, the primary signal for recovery mechanisms, was also responsible for signaling cell activation.

3.3 Eicosanoid Synthesis and Release

The eicosanoids are a group of biologically active molecules, derived from arachidonic acid, which have diverse effects on cells and are important mediators of pain and inflammation in vivo (SALMON and HIGGS 1987). These molecules are not stored in cells but are released as soon as they are synthesized. Prostaglandin synthesis and release from NC in vitro was the first stimulatory effect of the MAC to be described (SANDBERG et al. 1977). However, the relationship of prostaglandin release to cell death was not investigated at

Table 2. Activation of nucleated cells by the MAC

Cell	Mediators	Ca^{2+}-dependent	References
Neutrophils (R and H)	ROM	√	MORGAN and CAMPBELL 1985
	LT, TBX	√	SEEGER et al. 1986
Macrophages (H)	ROM	√	HANSCH et al. 1984; 1987
	PG	√	
	CK	√	
Glomerular mesangial cells (R)	ROM	?	ADLER et al. 1986;
	PG	?	LOVETT et al. 1987
	CK	?	
Glomerular epithelial cells (R)	PG	?	HANSCH et al. 1988
Rheumatoid synovial cells (H)	ROM	√	MORGAN et al. 1988b
	PG	√	DANIELS et al. 1990a, b
	LT	√	
	CK	?	VON KEMPIS et al. 1989
	Collagenase	?	JAHN et al. 1990
Pulmonary endothelial cells (H)	PG	√	SUTTORP et al. 1987a
Amniotic epithelial cells (H)	PG	?	ROONEY and MORGAN 1990b
Oligodendrocytes (R)	LT	?	SHIRAZI et al. 1987

R, rat; H, human; ROM, reactive oxygen metabolites; PG, prostaglandins; LT, leukotrienes; TBX, thromboxanes; CK, cytokines

that time. Recently, release of eicosanoids in response to nonlethal complement membrane attack has been observed in a variety of NC (Table 2). Two distinct biosynthetic pathways for eicosanoids exist in NCs: (1) the cyclooxygenase pathway which yields prostaglandins, prostacyclins, and thromboxanes and (2) the lipoxygenase pathway, which produces leukotrienes (SALMON and HIGGS 1987). The cyclooxygenase pathway is widely distributed in mammalian cells and is constitutively active, requiring only a supply of its substrate, arachidonic acid. The lipoxygenase pathway is restricted mainly to NC of hematogenous origin and is activated by increased $[Ca^{2+}]_i$. MAC-induced release of prostaglandins or leukotrienes has now been reported in a host of NC (Table 2). The role of calcium in signaling eicosanoid synthesis and release has not been investigated for all the NC listed in Table 2, but, in all cases in which calcium has been examined, it appears to be important in signaling the response. Chelation of extracellular calcium abolished MAC-induced eicosanoid release from most NC, and inhibitors of calmodulin activity have also been shown to block MAC-induced eicosanoid production (SUTTORP et al. 1987a), suggesting that the calcium signal is mediated via calmodulin. Unlike ROM release, which required formation of the complete MAC, eicosanoid release from some targets was observed at the C5b-8 stage, addition of C9 causing little or no enhancement of lysis. In other targets an absolute requirement for the complete MAC has been found (Table 2). The reasons for these inconsistencies are uncertain but may relate to different methods of cell preparation and formation of cell intermediates. In all cases, C5b-8 mediated stimulation of eicosanoid release was dependent on the presence of extracellular calcium, suggesting that the slow calcium leak known to occur through the C5b-8 lesion (RAMM et al. 1982; Morgan 1984) caused a sufficient increase in $[Ca^{2+}]_i$ to initiate release of eicosanoids (but not of ROM, see above) from these cells. MAC-induced release of leukotrience B_4 in rat oligodendrocyte × C6 glioma cell hybrids was calcium-dependent but was also inhibited by protein kinase antagonists (SHIRAZI et al. 1989). This result suggests that increased $[Ca^{2+}]_i$ activates protein kinases which in turn activate membrane phospholipases, thereby releasing arachidonic acid, the substrate for the lipoxygenase enzymes. The proposed pathway bears many similarities to that outlined above for cell recovery.

Due to the efficiency of recovery mechanisms, nonlethal complement membrane attack on NC in vitro using reactive lysis or cell intermediates and purified components is transient. Nevertheless, eicosanoid release often continues for many hours following attack (HANSCH et al. 1988; VON KEMPIS et al. 1989; DANIELS et al. 1990a, b). The signals responsible for this prolonged response to a transient insult have been studied in human rheumatoid synovial cells (DANIELS et al. 1990a, b). MAC-induced release of prostaglandin E_2 occurred in a biphasic manner, an early phase peaking at about 30 min after membrane attack followed by a second larger phase which persisted for more than 24 h. Removal of extracellular calcium during the initial attack abolished both phases of release, whereas inhibitors of protein synthesis abolished only the second prolonged phase. These results prompted the suggestion that the early phase was the result of activation (calcium-mediated) of preexisting enzymes of prostaglandin E_2 synthesis but that the second phase was due to de novo synthesis of cyclooxygenase enzymes (DANIELS et al. 1990a). The second phase was also inhibited by pertussis toxin, an inhibitor of G protein-mediated signaling systems (DANIELS et al. 1990b). This finding might imply a direct interaction of the MAC with receptors/G proteins in the membrane or could reflect an autocrine effect of the released prostaglandin E_2, acting via membrane receptors and G proteins, to further enhance cell stimulation (SMITH 1989). An involvement of cytokines in the mediation of these effects has also been demonstrated and will be detailed in a later section.

A variety of other pore-forming agents, including bacterial toxins (SUTTORP et al. 1987b), antibiotics (WIEGARD et al. 1988), melittin, calcium ionophores, and T cell perforins (DANIELS et al. 1990b) have also been shown to induce the release of eicosanoids from NC in the absence of cell death. In all cases, eicosanoid release was dependent on the presence of extracellular calcium, demonstrating that the influx of calcium through membrane pores is the primary stimulus.

3.4 Cytokine Release

The cytokines are a group of secreted proteins with diverse activities which act as local signals, either on the cell which secretes them (autocrine) or on nearby cells (paracrine). MAC-induced cytokine release was first observed in glomerular mesangial cells (LOVETT et al. 1987). Nonlethal membrane attack caused the release of an autocrine growth factor, resembling interleukin-1, which stimulated cell proliferation. Rheumatoid synovial cells exposed to nonlethal amounts of the MAC released interleukin-6 (DANIELS et al. 1990b). Cytokine release was not detectable until 5 h after attack but then rose steadily over the following 20 h. An autocrine role for the released interleukin-6 was suggested by the observation that neutralization of the cytokine with specific antibody inhibited release of Prostaglandin E_2 from these cells (DANIELS et al. 1990c). It is probable that the MAC mediates the release of cytokines from many cell types and this is likely to be a fruitful area of research in the future.

3.5 Other Nonlethal Effects of the MAC

Nonlethal effects of the MAC on specialized cellular functions have been observed in several NC types in vitro:

1. Membrane attack on glomerular epithelial cells caused enhanced synthesis of type IV collagen, a phenomenon implicated in the genesis of glomerulosclerosis (HANSCH et al. 1989; TORBOHM et al. 1990).
2. The MAC induced rheumatoid synovial cells to synthesize collagenase, an enzyme implicated in joint destruction (JAHN et al. 1990).
3. Expression of mRNA for myelin structural proteins in the glial cell responsible for myelin synthesis and maintenance, the oligodendrocyte, was inhibited by nonlethal membrane attack (SHIRAZI et al. 1990).

The important stimulatory effects of the MAC on platelets are discussed elsewhere in this volume (see Chap. 7).

3.6 Signaling of Nonlethal Effects of the MAC

The preceding sections describe some of the better studied stimulatory effects caused by nonlethal membrane attack on NC. The intracellular signals mediating these effects may differ in detail from cell to cell but the basic pathways are likely to be the same, and the weight of evidence suggests that the pathways mediating recovery also mediate cell activation. Recovery and activation are both dependent on calcium. Increased $[Ca^{2+}]_i$ caused by influx of calcium through the MAC pore initiates a chain of events which result in the activation of cellular enzyme systems and in the removal of MACs. The products of cell activation will vary from cell to cell, depending on the biosynthetic pathways available in the cell, but may include proinflammatory molecules, growth factors and tissue-degrading enzymes. The multiple signaling pathways implicated in MAC-induced cell activation, the complex interactions between the pathways, and the diverse products of activation are illustrated in Fig. 5.

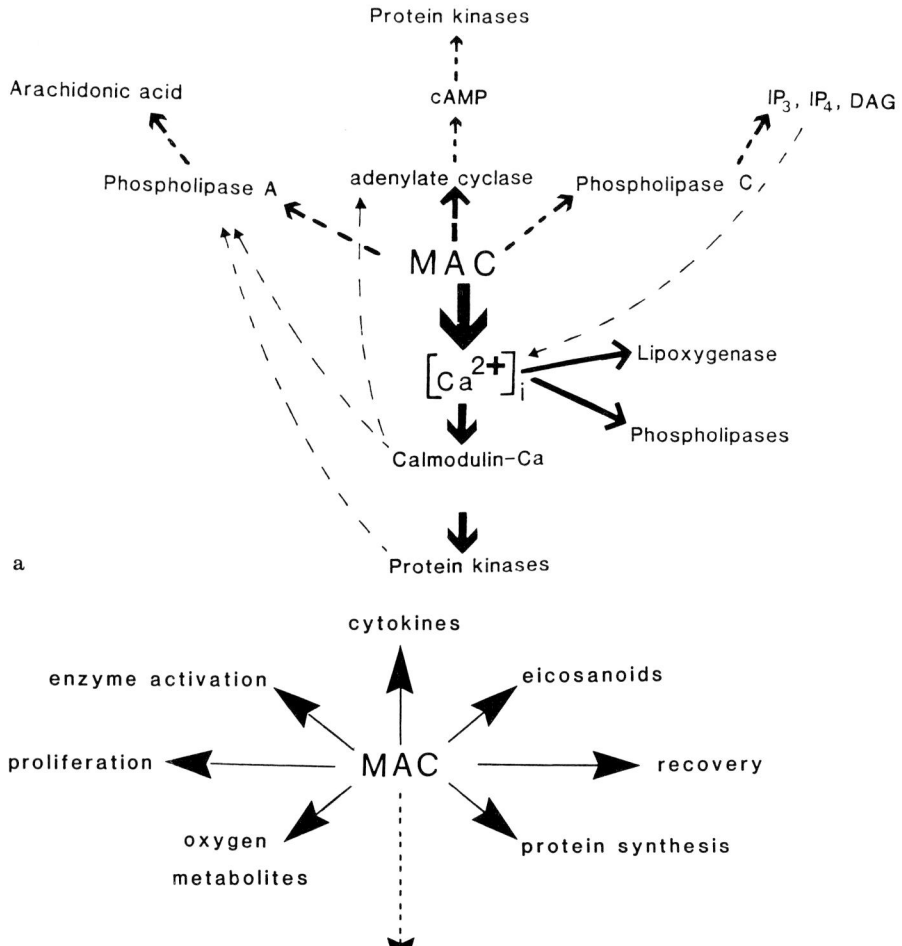

Fig. 5 a, b. Activation of nucleated cells by the MAC. **a** Intracellular signals for activation. Formation of the MAC on the membrane causes an increase in $[Ca^{2+}]i$ which, directly or indirectly, activates many cellular enzymes and results in the generation of secondary mediators. The MAC may also directly influence membrane signal generators thereby causing cell activation even in the absence of a rise in $[Ca^{2+}]i$. **b** The important consequences of complement membrane attack. The MAC activates diverse processes in the cell, initiates the release of potent mediators, and stimulates its own removal. Cytolysis is, in comparison to these nonlethal events, a rare consequence

4 Pathogenic Relevance of Cell Stimulation by the MAC

4.1 The MAC and Disease

Evidence of MAC formation provided by the measurement of fluid phase and tissue-associated complexes has implicated the MAC as a potential contributory factor in the pathogenesis of numerous inflammatory and immune-mediated diseases (reviewed in MORGAN 1989b, 1990 and summarized in Table 3). Despite widespread deposition of the potentially cytolytic MAC on tissues in these diseases, necrosis is not a dominant feature. Instead, the characteristic features of these conditions are those of cell activation and include proliferation and the release of inflammatory mediators. As detailed above, these are features which are typical of the effects of nonlethal membrane attack on NC in vitro. Although caution is essential when extrapolating results obtained from a simple in vitro system to the intact organism, the similarities are persuasive. Supportive evidence of the pathogenic relevance of nonlethal complement membrane attack has emerged from animal studies, in which it is possible to specifically

Table 3. Diseases in which the MAC is implicated

System	Disease	Evidence		
		TCC	MAC	Model
Renal	Lupus nephritis	√	√	√
	Post-streptococcal nephritis	?	√	√
	IgA nephropathy	?	√	√
	Membranous nephropathy	?	√	√
	MPGN Type III	?	√	—
Rheumatological	Rheumatoid arthritis	√	√	√
	Sjogren's syndrome	√	?	—
	Psoriatic arthritis	√	√	—
	Behcet's disease	√	√	—
Dermatological	Pemphigus vulgaris	?	√	—
	Bullous pemphigoid	?	√	—
	Dermatitis herpetiformis	?	√	—
Neurological	Multiple sclerosis	√	√	√
	Myasthenia gravis	?	√	√
	Cerebral lupus	√	?	—
Others	Atheroma/infarct	—	√	√
	Graves' disease	√	√	—
	Myositis	?	√	—

The table is derived in part from data in RAUTERBERG 1987 and MORGAN 1989a, b, 1990. TCC, fluid phase complexes detected; MAC, localization of MAC in tissue; model, evidence for MAC dependence from animal studies; √, evidence obtained; —, no evidence; ?, untested

block MAC formation by using animals genetically deficient in or immunochemically depleted of individual terminal complement components.

4.2 The MAC in Experimental Diseases

In the rat models of human nephritis, anti-glomerular basement membrane (GBM) nephritis and Heymann nephritis, the role of complement activation in the mediation of glomerular injury is well recognized (COUSER et al. 1985; SALANT et al. 1989). Involvement of the MAC in pathogenesis has recently been demonstrated in both these diseases. Rabbits deficient in C6 were refractory to the induction of anti-GBM nephritis whereas C6-sufficient animals rapidly developed disease (GROGGEL et al. 1985). In a variant of Heymann nephritis utilizing an isolated perfused rat kidney, proteinuria induced by perfusion of the antibody-sensitized kidney with serum did not occur if C8-depleted serum was substituted for whole serum in the perfusate (CYBULSKY et al. 1986).

The animal model for human myasthenia gravis, experimental autoimmune myasthenia gravis, is induced in rats by immunization with acetylcholine receptors (active) or infusion of antibodies against the acetylcholine receptor (passive). An involvement of complement was first indicated by the observation that complement depletion using cobra venom factor rendered animals resistant to the induction of disease (LENNON et al. 1978). Recently, an involvement of the MAC has been demonstrated (BIESECKER and GOMEZ 1989). Rats depleted in vivo of C6 by infusion of a specific antibody were refractory to the passive induction of disease whereas nondepleted animals rapidly developed symptoms.

4.3 The MAC in Human Diseases

There are many diseases listed in Table 3 and it is not my intention to suggest that the MAC is the major pathogenic factor in them all. My contention is simply that the MAC might contribute to a multifactorial pathogenesis in many diseases and that therapeutic strategies aimed at modulating the MAC might be beneficial in some of these. As an example of the correlations between the in vitro effects of the MAC on relevant cell types with pathogenic changes in the tissues, I will focus on one condition in which the MAC may contribute to disease pathogenesis.

There is abundant evidence for complement activation and MAC formation in the rheumatoid joint. Terminal complement complexes are present in the synovial fluid and the MAC is deposited in the synovial membrane (SANDERS et al. 1986; MORGAN et al. 1988a). Cells in the joint fluid and in the synovial membrane are therefore constantly exposed to the products of complement activation, including the MAC. Despite this chronic attack cell necrosis is not a major histological feature of the rheumatoid synovium which is thickened as a result of cellular infiltration and proliferation. The synovial fluid contains cytokines, eicosanoids, and reactive oxygen species, products of the resident and infiltrating cells. Synovial cells and phagocytic cells in vitro have been shown to produce all these inflammatory mediators in response to nonlethal complement membrane attack (Table 2). Membrane fragments bearing MACs have also been found in the synovial fluid, providing further evidence that nonlethal attack and cell recovery occur in vivo (MORGAN et al. 1988b). Although other factors also contribute to the pathogenesis of rheumatoid arthritis, there is ample reason to speculate that the MAC contributes to the initiation and/or perpetuation of inflammation in the rheumatoid joint. Therapies which inhibit the formation or effects of the MAC might therefore be of benefit in this disease and in numerous others in which nonlethal membrane attack contributes to pathology.

Acknowledgements. I thank the Wellcome Trust for financial support through the award of a Wellcome Senior Research Fellowship and Mrs. Jeannie Burt for secretarial assistance.

References

Adler S, Baker P, Johnson RJ, Ochi R, Pritzl P, Couser WG (1986) Complement membrane attack complex stimulates production of reactive oxygen metabolites by cultured rat mesangial cells. J Clin Invest 77: 762–767
Allan D, Michell RH (1979) The possible role of lipids in control of membrane fusion during secretion. Symp Soc Exp Biol 33: 323–336
Biesecker G, Gomez CM (1989) Inhibition of acute passive transfer experimental autoimmune myasthenia gravis with Fab antibody to complement C6. J Immunol 142: 2654–2659
Bordet J (1898) Sur l'agglutination et la dissolution des globules rouges par le serum d'aminaux injectes de sang defibrine. Ann Inst Pasteur 12: 688–695
Boyle MDP, Ohanian S, Borsos T (1976a) Lysis of tumor cells by antibody and complement VII. Complement-dependent ^{86}Rb release—a non-lethal event. J Immunol 117: 1346–1350
Boyle MDP, Ohanian S, Borsos T (1976b) Studies on the terminal stages of antibody-complement-mediated killing of a tumor cell. II. Inhibition of transformation of T* to dead cells by 3' 5' cAMP. J Immunol 116: 1276–1279
Boyle MDP, Ohanian S, Borsos T (1978) Effect of protease treatment on the sensitivity of tumor cells to antibody-GPC killing. Clin Immunol Immunopathol 10: 84–94
Campbell AK (1982) Intracellular calcium: its universal role as regulator. Wiley, Chichester, pp 257–304
Campbell AK, Luzio JP (1981) Intracellular calcium as a pathogen in cell damage initiated by the immune system. Experientia 37: 110–112
Campbell AK, Morgan BP (1985) Monoclonal antibodies demonstrate protection of polymorphonuclear leukocytes against complement attack. Nature 317: 164–166
Campbell AK, Daw RA, Luzio JP (1979) Rapid increase in intracellular free Ca^{2+} induced by antibody plus complement. FEBS Lett 107: 55–60
Campbell AK, Daw RA, Hallett MB, Luzio JP (1981) Direct measurement of the increase in intracellular free Ca^{2+} concentration in response to the action of complement. Biochem J 194: 551–560
Camussi G, Salvidio G, Biesecker G, Brentjens J, Andres G (1987) Heyman antibodies induce complement-dependent injury of rat glomerular visceral epithelial cells. J Immunol 139: 2906–2914
Carney DF, Koski CL, Shin ML (1985) Elimination of terminal complement intermediates from the plasma membrane of nucleated cells: the rate of disappearance differs for cells carrying C5b-7 or C5b-8 or a mixture of C5b-8 with a limited number of C5b-9. J Immunol 134: 1804–1809

Carney DF, Hammer CH, Shin ML (1986) Elimination of terminal complement complexes in the plasma membrane of nucleated cells: influence of extracellular Ca^{2+}. J Immunol 137: 263–270

Carney DF, Lang TI, Shin ML (1990) Multiple signal messengers generated by terminal complement complexes and their role in terminal complex elimination. J Immunol 145: 623–629

Casey PJ, Gilman AG (1988) G protein involvement in receptor-effector coupling. J Biol Chem 263: 2577–2580

Cheetham JJ, Chen RJB, Epand RM (1990) Interaction of calcium and cholesterol sulphate induces membrane destabilization and fusion: implications for the acrosome reaction. Biochim Biophys Acta 1024: 367–372

Cikes M (1970) Antigenic expression of an immune lymphoma during growth in vitro. Nature 225: 645–646

Cines DB, Schreiber AD (1979) Effect of anti-P1A1 antibody on human platelets. 1. The role of complement. Blood 53: 567–577

Colbran RJ, Schworer CM, Hasimoto Y, Fong Y-L, Rich DP, Smith KM, Soderling TR (1989) Calcium/calmodulin dependent protein kinase II. Biochem J 258: 313–325

Cooper NR, Polley MJ, Oldstone MBA (1974) Failure of terminal complement components to induce lysis of Moloney virus transformed lymphocytes. J Immunol 112: 866–868

Couser WG, Baker PJ, Adler S (1985) Complement and the direct mediation of immune glomerular injury: a new perspective. Kidney Int 28: 879–890

Cox JA (1988) Interactive properties of calmodulin. Biochem J 249: 621–629

Cybulsky AV, Rennke HG, Feintzeig ID, Salant DJ (1986) Complement-induced glomerular epithelial cell injury. Role of the membrane attack complex in rat membranous nephropathy. J Clin Invest 77: 1096–1107

Cybulsky AV, Salant D, Quigg RJ, Badalamenti J, Bonventre JV (1989) Complement C5b-9 complex activates phospholipases in glomerular epithelial cells. Am J Physiol 257: F826–F836

Cybulsky AV, Bonventre JV, Quigg RJ, Lieberthal W, Salant DJ (1990) Cytosolic calcium and protein kinase C reduce complement-mediated glomerular epithelial injury. Kidney Int 38: 803–811

Daniels RH, Houston WA, Petersen MM, Williams JD, Williams BD, Morgan BP (1990a) Stimulation of rheumatoid synovial cells by non-lethal complement membrane attack. Immunology 69: 237–242

Daniels RH, Williams BD, Morgan BP (1990b) Human rheumatoid synovial cell stimulation by the membrane attack complex and other pore-forming toxins in vitro; the role of calcium in cell activation. Immunology 71: 312–316

Daniels RH, Williams BD, Morgan BP (1990c) Non-lethal complement membrane attack on cultured synovial cells induces G-protein and calcium-dependent PGE_2 release and release of IL-6. Compl Inflamm 7: 137 (abstract)

Ehrlich P, Morgenroth J (1899) Zur Theorie der Lysinwirkung. Berl Klin Wochenschr 36: 6–9

Ferrone S, Cooper NR, Pellegrino MA, Reisfeld RA (1973) Interaction of histocompatibility (HLA) antibodies and complement with synchronized human lymphoid cells in continuous culture. J Exp Med 137: 55–68

Fischelson Z, Kopf E, Paas Y, Ross L, Reiter Y (1989) Protein phosphorylation as a mechanism of resistance against complement damage. Prog Immunol 7: 205–210

Goldberg B, Green H (1959) The cytotoxic action of immune gamma globulin and complement on Krebs ascites tumor cells. I. Ultrastructural studies. J Exp Med 109: 505–510

Green H, Goldberg B (1960) The action of antibody and complement on mammalian cells. Ann NY Acad Sci 87: 352–361

Green H, Barrow P, Goldberg B (1959) Effect of antibody and complement on permeability control in ascites tumor cells and erythrocytes. J Exp Med 110: 689–713

Groggel GC, Salant DJ, Darby C, Rennke HG, Couser WG (1985) Role of terminal complement pathway in the heterologous phase of antiglomerular basement membrane nephritis. Kidney Int 27: 643–651

Hallett MB, Luzio JP, Campbell AK (1981) Stimulation of Ca^{2+}-dependent chemiluminescence in rat polymorphonuclear leucocytes by polystyrene beads, and the non-lytic action of complement. Immunology 44: 569–577

Halliwell B, Gutteridge JMC (1984) Oxygen toxicity, oxygen radicals, transition metals and disease. Biochem J 219: 1–14

Hansch GM (1988) The homologous species restriction of complement attack: structure and function of the C8-binding protein. In: Podack ER (ed) Cytotoxic effector mechanisms. Springer, Berlin Heidelberg New York, pp 109–118 (Current topics in microbiology and immunology, vol 140)

Hansch GM, Seitz M, Marinotti G, Betz M, Rauterberg EW, Shin ML (1984) Macrophages release arachidonic acid, prostaglandin E_2, and thromboxane in response to late complement components. J Immunol 133: 2145–2150

Hansch GM, Seitz M, Betz M (1987) Effect of the late complement components C5b-9 on human monocytes: release of prostanoids, oxygen radicals and of a factor inducing cell proliferation. Int Arch Allergy Appl Immunol 82: 317–320

Hansch GM, Betz M, Gunther J, Rother KO, Sterzel B (1988) The complement membrane attack complex stimulates the prostanoid production of cultured glomerular mesangial cells. Int Arch Allergy Appl Immunol 85: 87–93

Hansch GM, Torbohm I, Rother K (1989) Chronic glomerulonephritis: inflammatory mediators stimulate the collagen synthesis in glomerular epithelial cells. Int Arch Allergy Appl Immunol 88: 139–143

Imagawa DK, Barbour SE, Morgan BP, Wright TM, Hin HS, Ramm LE (1987) Role of complement C9 and calcium in the generation of arachidonic acid metabolites from rat polymorphonuclear leukocytes. Mol Immunol 24: 1263–1271

Jackson MB, Stephens CL, Lecar H (1981) Single channel currents induced by complement in antibody-coated cell membranes. Proc Natl Acad Sci USA 79: 6421–6425

Jahn B, von Kempis J, Hansch GM (1990) Induction of prostaglandin E2 (PGE2) and collagenase synthesis in human synovial fibroblast-like cells (SFC) by terminal complement components C5b-9. Compl Inflamm 7: 138 (abstract)

Kaliner M, Austen KF (1974) Adenosine 3' 5'-monophosphate: inhibition of complement mediated cell lysis. Science 183: 659–661

Kerjaschki D, Schulze M, Binder S, Kain R, Ojha PP, Susani M, Horvat R, Baker PJ, Couser WG (1989) Transcellular transport and membrane insertion of the C5b-9 membrane attack complex of complement by glomerular epithelial cells in experimental membranous nephropathy. J Immunol 143: 546–552

Koski CL, Ramm LE, Hammer CH, Mayer MM, Shin ML (1983) Cytolysis of nucleated cells by complement. Cell death displays multi-hit characteristics. Proc Natl Acad Sci USA 80: 3816–3820

Lachmann PJ, Coombs RR, Fell HB (1969) The breakdown of embryonic (chick) cartilage and bone cultivated in the presence of complement-sufficient antiserum. 3. Immunological analysis. Int Arch Allergy Appl Immunol 36: 469–485

Lennon VA, Seybold ME, Lindstrom JM, Cochrane C, Ulevitch R (1978) Role of complement in the pathogenesis of experimental autoimmune myasthenia gravis. J Exp Med 147: 973–983

Levine YK (1972) Physical studies of membrane structure. Prog Biophys Mol Biol 24: 1–74

Linscott WD (1970) An antigen density effect on the hemolytic efficiency of complement. J Immunol 104: 1307–1309

Lo TN, Boyle MDP (1979) Relationship between the intracellular cyclic adenosine 3':5'-monophosphate level of tumor cells and their sensitivity to killing by antibody and complement. Cancer Res 39: 3156–3162

Lovett DH, Haensch GM, Goppelt M, Resch K, Gemsa F (1987) Activation of glomerular mesangial cells by the terminal membrane attack complex of complement. J Immunol 138: 2473–2480

Mayer MM (1961) Development of a one-hit theory of immune hemolysis. In: Heidelbeger M, Plescia DJ (eds) Immunochemical approaches to problems in microbiology. Rutgers, New Jersey, pp 268–279

Mayer MM (1972) Mechanisms of cytolysis by complement. Proc Natl Acad Sci USA 69: 2954–2958

Mayer MM, Imagawa DK, Ramm LE, Whitlow MB (1983) Membrane attack by complement and its consequences. Prog Immunol 5: 427–444

Meri S, Morgan BP, Davies A, Daniels RH, Olavesen MG, Waldmann H, Lachmann PJ (1990) Human protectin (CD59), an 18 000–20 000 MW complement lysis restricting factor, inhibits C5b-8 catalysed insertion of C9 into bilayers. Immunology 71: 1 9

Moller E, Moller G (1962) Quantitative studies of sensitivity of normal and neoplastic cells to the cytotoxic action of isoantibodies. J Exp Med 115: 27–534

Morgan BP (1984) The biochemistry and pathology of complement component C9. PhD thesis, University of Wales, pp 235–274

Morgan BP (1988) Non-lethal complement membrane attack on human neutrophils: transient cell swelling and metabolic depletion. Immunology 63: 71–77

Morgan BP (1989a) Mechanisms of tissue damage by the membrane attack complex of complement. Compl Inflamm 6: 104–111

Morgan BP (1989b) Complement membrane attack on nucleated cells: resistance, recovery and non-lethal effects. Biochem J 264: 1–14

Morgan BP (1990) Complement, clinical aspects and relevance to disease. Academic, London

Morgan BP, Campbell AK (1985) The recovery of human polymorphonuclear leucocytes from sublytic complement attack is mediated by changes in intracellular free calcium. Biochem J 231: 205–208

Morgan BP, Campbell AK, Luzio JP, Hallett MB (1984) Recovery of polymorphonuclear leucocytes from complement attack. Biochem Soc Trans 12: 779–780

Morgan BP, Imagawa DK, Dankert JR, Ramm LE (1986a) Complement lysis of U937, a nucleated mammalian cell line in the absence of C9: effect of C9 on C5b-8 mediated cell lysis. J Immunol 136: 3402–3406

Morgan BP, Luzio JP, Campbell AK (1986b) Intracellular Ca^{2+} and cell injury: a paradoxical role of Ca^{2+} in complement membrane attack. Cell Calcium 7: 399–411

Morgan BP, Dankert JR, Esser AF (1987) Recovery of human neutrophils from complement attack: removal of the membrane attack complex by endocytosis and exocytosis. J Immunol 138: 246–253

Morgan BP, Daniels RH, Williams BD (1988a) Measurement of terminal complement complexes in rheumatoid arthritis. Clin Exp Immunol 73: 473–478

Morgan BP, Daniels RH, Watts MJ, Williams BD (1988b) In vivo and in vitro evidence of cell recovery from complement attack in rheumatoid synovium. Clin Exp Immunol 73: 467–472

Nose M, Katoh M, Okada N, Kyogoku M, Okada H (1990) Tissue distribution of HRF20, a novel factor preventing the membrane attack of homologous complement, and its prediomiant expression on endothelial cells in vivo. Immunology 70: 145–149

Nuttal G (1888) Experimente über die Bakterienfeindlichen Einflüsse des tierischen Körpers. Z Hyg Infektionskr 4: 353–356

Ohanian SH, Borsos T (1975) Lysis of tumor cells by antibody and complement. II. Lack of correlation between amount of C4 and C3 fixed and cell lysis. J Immunol 114: 1292–1295

Ohanian SH, Schlager SI (1981) Humoral immune killing of nucleated cells: mechanisms of complement-mediated attack and target cell defense. CRC Crit Rev Immunol 1: 165–209

Ohanian SH, Schlager SI, Borsos T (1977) Molecular interactions of cells with antibody and complement. influence of metabolic and physical properties of the target on the outcome of humoral immune attack. Contemp Top Mol Immunol 7: 153–175

Ohanian SH, Schlager SI, Yamazaki M, Ishida B (1979) Effect of specific phospholipids on the antibody-complement killing of nucleated cells. J Immunol 123: 1014–1019

Ohanian SH, Schlager SI, Saha S (1982) Effect of lipids, structural precursors of lipids and fatty acids on complement-mediated killing on antibody-sensitized nucleated cells. Mol Immunol 19: 535–542

Podack ER, Muller-Eberhard HJ (1981) Complement-mediated membrane injury of tumor cells: release of membrane vesicles. Fed Proc 40: 359 (abstract)

Pellegrino MA, Ferrone S, Cooper NR, Dierich MP, Reisenfeld RA (1974) Variation in susceptibility of a human lymphoid cell line to immune lysis during the cell cycle: lack of correlation with antigen density and complement binding. J Exp Med 140: 578–590

Raisz LG, Sandberg AL, Goodson JM, Simmons HA, Mergenhagen SE (1974) Complement-dependent stimulation of prostaglandin synthesis and bone resorption. Science 185: 787–791

Ramm LE, Mayer MM (1980) Life span and size of the transmembrane channels formed by large doses of complement. J Immunol 124: 2281–2287

Ramm LE, Whitlow MB, Mayer MM (1982) Size of the transmembrane channels produced by complement proteins C5b-8. J Immunol 129: 1143–1146

Ramm LE, Whitlow MB, Mayer MM (1983a) Size distribution and stability of the transmembrane channels formed by complement complex C5b-9. Mol Immunol 20: 155–160

Ramm LE, Whitlow MB, Koski CL, Shin ML, Mayer MM (1983b) Elimination of complement channels from the plasma membranes of U937, a nucleated mammalian cell line: temperature dependence of the elimination rate. J Immunol 131: 1411–1415

Ramm LE, Whitlow MB, Mayer MM (1984) Complement lysis of nucleated cells: effect of temperature and puromycin on the number of channels required for cytolysis. Mol Immunol 21: 1015–1019

Rauterberg EW (1987) Demonstration of complement deposits in tissues. In: Rother K, Till GO (eds) The complement system. Springer, Berlin Heidelberg New York, pp 287–326

Richardson PJ, Luzio JP (1980) Complement-mediated production of plasma membrane vesicles from rat fat cells. Biochem J 186: 897–906

Roberts PA, Morgan BP, Campbell AK (1985) 2-chloroadenosine inhibits complement-induced reactive oxygen metabolite production and recovery of human polymorphonuclear leucocytes attacked by complement. Biochim Biophys Res Commun 126: 692–697

Rooney IA, Morgan BP (1990a) Protection of human amniotic epithelial cells (HAEC) from complement-mediated of three complement inhibitory membrane proteins. Immunology 71: 308–311

Rooney IA, Morgan BP (1990b) Non-lethal doses of antibody and complement stimulate release of prostaglandin E_2 from human amniotic cells in vitro. Biochem Soc Trans 18: 617

Rooney IA, Davies A, Griffiths D, Williams JD, Davies M, Meri S, Lachmann PJ, Morgan BP (1991) The complement inhibiting protein, protectin (CD59 antigen) is present and functionally active on glomerular epithelial cells. Clin Exp Immunol 83: 251–256

Salant DJ, Quigg RJ, Cybulsky AV (1989) Heymann nephritis: mechanisms of renal injury. Kidney Int 35: 976–990

Salmon JA, Higgs GA (1987) Prostaglandins and leukotrienes as inflammatory mediators. Br Med Bull 43: 285–296

Sandberg AL, Raisz LG, Goodson MJ, Simmons HA, Mergenhagen SE (1977) Limitation of bone resorption by the classical and alternative pathways and its mediation by prostaglandin. J Immunol 119: 1378–1381

Sanders ME, Kopicky JA, Wigley FM, Shin ML, Frank MM, Joiner KA (1986) Membrane attack complex of complement in rheumatoid synovial tissue demonstrated by immunofluorescent microscopy. J Rheumatol 13: 1028–1034

Schlager SI, Ohanian SH, Borsos T (1976) Inhibition of antibody-complement mediated killing of tumor cells by hormones. Cancer Res 36: 3672–3677

Scolding NJ, Houston WAJ, Morgan BP, Campbell AK, Compston DAS (1989a) Reversible injury of cultured rat oligodendrocytes by complement. Immunology 67: 441–446

Scolding NJ, Morgan BP, Houston WAJ, Linington C, Campbell AK, Compston DAS (1989b) Vesicular removal by oligodendrocytes of membrane attack complexes formed by activated complement. Nature 339: 620–622

Scolding NJ, Morgan BP, Frith S, Campbell AK, Compston DAS (1990) Intracellular calcium and oligodendrocyte injury. J Neurol Neurosurg Psychiatry 53: 811 (abstract)

Seeger W, Suttorp N, Helliwig A, Bhakdi S (1986) Noncytolytic complement complexes may serve as calcium gates to elicit leukotriene B4 generation in human polymorphonuclear leukocytes. J Immunol 137: 1286–1293

Segerling MS, Ohanian SH, Borsos T (1975a) Chemotherapeutic drugs increase killing of tumor cells by antibody and complement. Science 188: 55–57

Segerling MS, Ohanian SH, Borsos T (1975b) Enhancing effect by metabolic inhibitors on the killing of tumour cells by antibody and complement. Cancer Res 35: 3195–3203

Shin ML, Paznekas WA, Mayer MM (1978) On the mechanism of membrane damage by complement: the effect of length and saturation of the acyl chains in liposomal bilayers and the effect of cholesterol concentration in sheep erythrocytes and liposomal membranes. J Immunol 120: 1996–2002

Shirazi Y, Imagawa DK, Shin ML (1987) Release of leukotriene B4 from sublethally injured oligodendrocytes by terminal complement complexes. J Neurochem 48: 271–278

Shirazi Y, McMorris FA, Shin ML (1989) Arachidonic acid mobilization and phosphoinositide turnover by the terminal complement complex, C5b-9, in rat oligodendrocyte × C6 glioma cell hybrids. J Immunol 142: 4385–4391

Shirazi Y, Macklin WB, Shin ML (1990) Terminal complement complexes (TCC) inhibit myelin protein mRNA expression in oligodendrocytes (OLG). FASEB J 4: A2017 (abstract)

Sims PJ, Weidmer T (1986) Repolarization of the membrane potential of blood platelets after complement damage: evidence for a Ca^{2+}-dependent exocytic elimination of C5b-9 pores. Blood 68: 556–561

Slater TF (1984) Free-radical mechanisms in tissue injury. Biochem J 222: 1–15

Smith WL (1989) The eicosanoids and their biochemical mechanisms of action. Biochem J 259: 315–324

Stein JM, Luzio JP (1990) Ectocytosis caused by sublytic autologous complement attack on human neutrophils: the sorting of endogenous plasma membrane proteins and lipids into shed vesicles. Biochem J 274: 381–386

Stephens CL, Henkart PA (1979) Electrical measurements of complement-mediated membrane damage in cultured nerve and muscle cells. J Immunol 122: 455–458

Suttorp N, Seeger W, Zinsky S, Bhakdi S (1987a) Complement complex C5b-8 induces PGI_2 formation in cultured endothelial cells. Am J Physiol 253: C13–21

Suttorp N, Seeger W, Zucker-Reimann J, Roka L, Bhakdi S (1987b) Mechanism of leukotriene generation in polymorphonuclear leukocytes by staphylococcal alpha-toxin. Infect Immun 55: 104–110

Taylor CW (1990) The role of G proteins in transmembrane signalling. Biochem J 272: 1–13

Thielens NM, Lohner K, Esser AF (1988) Human complement component C9 is a calcium binding protein: structural and functional implications. J Biol Chem 263: 6665–6670

Torbohm I, Schonermark M, Wingen AM, Berger B, Rother K, Hansch GM (1990) C5b-8 and C5b-9 modulate the collagen release of human glomerular epithelial cells. Kidney Int 37: 1098–1104

Von Kempis J, Torbohm I, Schonermark M, Jahn B, Seitz M, Hansch GM (1989) Effect of the late complement components C5b-9 and of platelet-derived growth factor on the prostaglandin release of human synovial fibroblast-like cells. Int Arch Allergy Appl Immunol 90: 248–255

Wiegard R, Betz M, Hansch GM (1988) Nystatin stimulates prostaglandin E synthesis and formation of diacylglycerol in human monocytes. Agents Actions 24: 243–250

Yoo TJ, Chin HC, Spector AA, Whitaker RJ, Denning CM, Lee NF (1980) Effect of fatty acid modifications of cultured hepatoma cells on susceptibility to complement-mediated cytolysis. Cancer Res 40: 1084–1090

Glycosyl-Phosphatidylinositol Anchoring of Membrane Proteins

D. M. LUBLIN

1 Introduction	141
2 Structure of GPI Anchors	142
2.1 Biochemical Structure of GPI Anchors	142
2.2 Biosynthesis of GPI-Anchored Membrane Proteins	144
2.3 Peptide Signal Sequence for GPI Anchor Attachment	145
2.4 Paroxysmal Nocturnal Hemoglobinuria and Other Defects in the Pathway for GPI Anchoring	148
3 Functions of GPI Anchors	149
3.1 Membrane Attachment and Release	150
3.2 Protein Lateral Mobility	151
3.3 Cell Activation	153
3.4 Sorting in Polarized Epithelial Cells	155
4 Conclusions	156
References	157

1 Introduction

Three of the five complement regulatory proteins discussed in detail in other chapters of this volume are anchored to the plasma membrane through a glycosyl-phosphatidylinositol (GPI) structure; specifically, decay accelerating factor (DAF), membrane inhibitor of reactive lysis (MIRL), and homologous restriction factor (HRF). This GPI form of membrane anchoring has gone from being an unusual observation of the late 1970s relating to the release of a few hydrolytic enzymes (alkaline phosphatase, acetylcholinesterase, and 5'-nucleotidase) from plasma membranes by PI-specific phospholipase C (PI–PLC) to recognition as a fairly common mode of membrane attachment (LOW 1987). This is evidenced with almost every new issue of a journal and is reflected in the common use of this mode of attachment by cell membrane proteins, including complement regulatory proteins and other immunologically important molecules, e.g., Thy-1, one form of lymphocyte function-associated antigen 3 (LFA-3), and one form of IgG F_c receptor III.

Division of Laboratory Medicine, Departments of Pathology and Medicine, Washington University School of Medicine, St. Louis, MO 63110, USA

After some time in relative obscurity, GPI-anchored membrane proteins have become an area of intense investigation in recent years. Almost all of the initial work was focused on the structure and biosynthesis of these molecules. Only the last several years have seen direct evidence for the function of this class of membrane proteins. With several excellent reviews on GPI-anchored membrane proteins already available (Low 1987; Low and Saltiel 1988; Ferguson and Williams 1988; Cross 1990), we will instead pay particular attention to functional issues in this chapter. Additionally, as this volume concerns the complement regulatory proteins, we will take every opportunity to use these as illustrative examples where applicable. This refers especially to DAF, since it is one of the best studied of the GPI-anchored membrane proteins.

2 Structure of GPI Anchors

GPI-anchored membrane proteins were initially recognized because they could be removed from the cell surface by PI–PLC. For many proteins in the growing list of GPI-anchored membrane proteins, this is still the only evidence that they utilize a GPI anchor; however, this form of data has always held up when more structural data has become available. A detailed biochemical picture of GPI anchors was produced as a result of the work of many investigators, including Ferguson, Cross, Hart, Englund and their colleagues (see reviews cited above for details), on the structure of the GPI anchor of the trypanosome variant surface glycoprotein (VSG), eventually culminating in determination of the complete chemical structure (Ferguson et al. 1988). VSG represents an excellent sample for this analysis because it comprises 10% of the protein of the parasite. Parallel work on other GPI anchors, particularly rat Thy-1 (Homans et al. 1988) and human erythrocyte acetylcholinesterase (AChE) (Roberts et al. 1988a, b), has extended this work to mammalian proteins. Further data has elucidated some of the biochemical steps involved in producing these structures. The present picture of the biosynthesis and structure of GPI anchors was built on many years of work and many individual steps; here we will simply present the structure as it is now known.

2.1 Biochemical Structure of GPI Anchors

The biochemical structures of trypanosome VSG (Ferguson et al. 1988), rat brain Thy-1 (Homans et al. 1988), and human erythrocyte AChE (Roberts et al. 1988a, b) are shown in Fig 1. The complete structure of the first two anchors was deduced by a combination of nuclear magnetic resonance spectroscopy, mass spectrometry, and chemical and enzymatic digestions, whereas the structure of the latter was based on fast atom bombardment mass spectrometry. One of the most

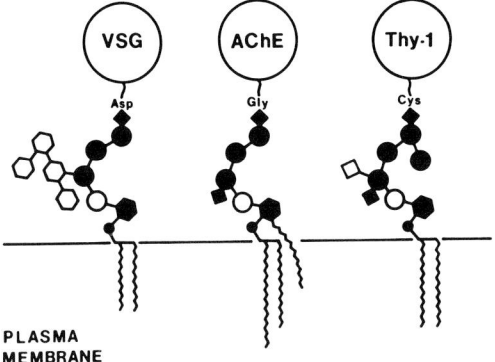

Fig. 1. The structure of the GPI anchors of trypanosome VSG (FERGUSON et al. 1988), rat Thy-1 (HOMANS et al. 1988), and human acetylcholinesterase (AChe) (ROBERTS et al. 1988a, b). Phosphoethanolamine (◆), mannose (●), glucosamine (○), N-acetylgalactosamine (◇), galactose (⬡), inositol (⬢), and phosphate (●); long-chain fatty acyl or alkyl groups shown as wavy lines. This diagram highlights the conservation of the core ethanolamine–phosphate–mannose$_3$–glucosamine–inositol structure and illustrates the differences in side chains and fatty acids (see text for additional details). (From Low 1989)

striking observations is the remarkable conservation of this GPI anchor throughout the large evolutionary distance from protozoa to mammals. The backbone is absolutely conserved among these organisms: the α-COOH group of the COOH-terminal amino acid of the protein is in amide linkage to an ethanolamine–P–6–Man–α1–2–Man–α1–6–Man–α1–4–GlcN–α1–6–inositol phospholipid. The phospholipid moiety is inserted in the outer leaflet of the plasma membrane bilayer and forms the actual attachment of the protein to the cell.

In contrast to the remarkable conservation of the backbone structure of the GPI anchor, there are several differences in the side chains. In the case of VSG there is a heterogeneous (0–8 residue) branched galactose side chain attached to the mannose residue (linked to glucosamine), whereas for both mammalian anchors there is an ethanolamine phosphate group attached to that residue. Based on compositional data showing 2 mol/mol ethanolamine in bovine and human AChE (ROBERTS et al. 1987; HAAS et al. 1986), human placental alkaline phosphatase (OGATA et al. 1988), human DAF (MEDOF et al. 1986), hamster brain scrapie prion protein (STAHL et al. 1987), and squid glycoprotein-2 (WILLIAMS et al. 1988), the ethanolamine side chain might be a general feature or GPI anchors in higher eukaryotes. Rat brain Thy-1 also contains an N-acetylgalactosamine residue and a fourth mannose residue, but these might represent tissue- or species-specific differences (the extra mannose residue in rat brain Thy-1 is absent from rat thymocyte Thy-1; TSE et al. 1985).

An additional source of heterogeneity between GPI anchors resides in the fatty acids, which might represent protein, tissue, or species variation. In VSG the fatty acids are in dimyristylglycerol (FERGUSON et al. 1985), whereas in all the GPI anchors from higher organisms the fatty acids are more heterogeneous and in many cases comprise a 1-O-alkyl-2-O-acylglycerol group rather than a 1,2-O-diacylglycerol. A well studied example is human AChE (ROBERTS et al. 1988a, b), in which the 1-O-alkyl group is a mixture of 18:0 and 18:1 and the 2-O-acyl group comprises 18:1, 22:4, and 22:5. It is interesting to note that the dimyristylglycerol moiety in VSG actually arises from a novel process of fatty acid remodeling (MASTERSON et al. 1990); the initial GPI anchor is synthesized with fatty acids that are more hydrophobic than myristate and then subsequently replaced by deacylation and reacylation with myristate at both positions. This appears to be a process particular to the trypanosome and not seen in higher eukaryotes, and the physiological reason for such a remodeling is not known.

Another fatty acid modification is the acylation of the inositol ring. Palmitoylation of the inositol at the 2-position has been demonstrated for human AChE (ROBERTS et al. 1988b) as shown in Fig. 1. Similar modification occurs for VSG (MAYOR et al. 1990b) and for human DAF (WALTER et al. 1990). It is not clear exactly where in the biosynthetic pathway this acylation occurs or even whether the GPI anchors with and without acylated inositol rings are directly related. This acylation results in resistance to cleavage by PI–PLC (ROBERTS et al. 1988b), raising the possibility of a physiological role for this modification (see below).

2.2 Biosynthesis of GPI-Anchored Membrane Proteins

The overall biosynthesis of GPI-anchored membrane proteins involves synthesis of a GPI anchor precursor, synthesis of the protein with a cleavable, hydrophobic COOH-terminal signal sequence (and a cleavable NH_2-terminal leader peptide for translocation into the endoplasmic reticulum), and replacement of the COOH-terminal peptide with the preformed GPI anchor in the endoplasmic reticulum. The initial finding was that the cDNA for the protein encodes 17–31 amino acids (with a hydrophobic stretch) at the COOH-terminal that are not present in the mature protein. This was discovered first for VSG (BOOTHROYD et al. 1981) and subsequently noted in all GPI-anchored proteins that have had direct peptide and cDNA sequences determined (for list see FERGUSON and WILLIAMS 1988; CROSS 1990). The actual nature of the signal sequence for GPI anchor attachment is discussed below.

The rapidity of addition of the GPI anchor to proteins (radiolabeled myristate is incorporated into VSG within 1 min; BANGS et al. 1985; FERGUSON et al. 1986) suggested that this posttranslational modification takes place in the endoplasmic reticulum. This has been confirmed by the finding that GPI anchors are still added to a membrane glycoprotein in a yeast mutant (sec 18) in which transport from the endoplasmic reticulum to the Golgi apparatus is blocked (CONZELMANN et al. 1988a). Additional evidence comes from the finding that GPI anchors can be added to Thy-1 and DAF in an in vitro translation system when canine pancreatic microsomal membranes are added (FASEL et al. 1989).

Work on VSG from the laboratories of HART and ENGLUND (KRAKOW et al. 1986; MASTERSON et al. 1989, 1990; DOERING et al. 1989) and of CROSS (MENON et al. 1988, 1990a,b; MAYOR et al. 1990a,b) has identified a GPI anchor precursor and putative biosynthetic intermediates. This work was made possible in part by the development of a trypanosome cell-free system for synthesis of GPI anchor precursors (MASTERSON et al. 1989; MENON et al. 1990b). The structural characterization of these precursors suggests a pathway for biosynthesis of the GPI anchor. As might be suggested based on the anchor backbone structure (Fig. 1), the anchor is assembled by the sequential glycosylation of PI.

The first step in the pathway is the transfer of GlcNAc from UDP-GlcNAc to PI to form GlcNAc-PI. A deacetylation step yields GlcN-PI, which is then sequentially mannosylated from a dolichol-P-mannose donor (MENON et al. 1990a) to produce Man_3-GlcN-PI. The addition of ethanolamine phosphate and the remodeling of the fatty acids to myristate (MASTERSON et al. 1990) produce a precursor designated glycolipid A (KRAKOW et al. 1986) or P2 (MENON et al. 1988); this is identical to the GPI anchor of VSG except for the free amino group of ethanolamine at one end (which will form an amide bond to the protein) and the absence of the galactosyl side chain, added as a late modification probably in the Golgi apparatus (BANGS et al. 1985, 1988). Another precursor, glycolipid C (KRAKOW et al. 1986) or P3 (MENON et al. 1988), was also found; it differed from glycolipid A or P2 in its resistance to cleavage by PI–PLC. Analysis of P2 and P3 demonstrated that P3 differed only in having an acylated inositol ring (MAYOR et al. 1990b) similar to the modification described for human AChE (Fig. 1) (ROBERTS et al. 1988a). It is not clear whether the synthesis of glycolipid A/P2 and glycolipid C/P3 proceeds in parallel or whether these species are interconverted at some point by acylation or deacylation. Furthermore, the role of the acylated form is unknown; it does protect the anchor from PI–PLC, and it might function in this regard during biosynthesis of the anchor, protecting it from cytoplasmic PI–PLC (MAYOR et al. 1990b). Susceptibility or resistance to PI–PLC varies in a manner that is cell- (Low et al. 1988) and species- specific (ROBERTS et al. 1987) so this might be a method of regulating the expression or function of the protein.

2.3 Peptide Signal Sequence for GPI Anchor Attachment

The preformed GPI anchor (whose structure and biosynthesis are described above) is added to the protein as an early posttranslational modification in the endoplasmic reticulum. This probably occurs via a transamidation reaction, in which an internal peptide bond is broken and the α-COOH group forms an amide bond with ethanolamine from the GPI anchor. The signal for anchor attachment has been mapped to the COOH-terminal of the protein (CARAS et al. 1987; TYKOCINSKI et al. 1988; WANECK et al. 1988a). For example, the 37 COOH-terminal amino acids of DAF can convert herpes simplex virus type 1 glycoprotein D into a GPI-anchored protein (CARAS et al. 1987). However, inspection of the COOH-terminals of known GPI-anchored proteins does not provide any clear consensus sequence serving as a signal (see Table 1 in CROSS 1990; Table 3 in FERGUSON and WILLIAMS 1988). The only real pattern seen in GPI-anchored proteins is that the COOH-terminal is hydrophobic and lacks a cytoplasmic domain. The hydrophobic stretch of amino acids often contains hydrophilic or even charged residues, leading to the early suggestion that the consensus signal for GPI anchoring is a "weakly hydrophobic" domain without a cytoplasmic tail (WANECK et al. 1988b). This hypothesis was supported by the finding that conversion of a single Asp residue to Val in the hydrophobic domain of Qa-2 changed that protein from GPI-anchored to transmembrane (WANECK et al. 1988b). However, additional studies did not support this initial hypothesis, because either the hydrophobic NH_2-terminal signal peptide from human growth hormone, a random hydrophobic sequence (CARAS and WEDDELL 1989), or the strongly hydrophobic transmembrane domain of membrane cofactor protein (MCP) (SMITH et al. 1990) could substitute for the moderately hydrophobic domain in DAF and still yield GPI anchoring. Furthermore, although addition of a cytoplasmic domain to a GPI-anchored protein prevents GPI anchoring in some cases (BERGER et al. 1989; ULKER et al. 1990), in other cases it does not (SU and BOTHWELL 1989; SMITH et al. 1990). So neither the "weakly" hydrophobic nature of the COOH-terminal domain or the lack of a cytoplasmic tail are absolute requirements for GPI anchoring, although for some specific GPI-anchored proteins these factors seem to be involved.

To get at the nature of the consensus signal for GPI anchoring, mutations have been introduced into the COOH-terminal region that is the signal for GPI anchoring. Detailed studies have been done for DAF (CARAS and WEDDELL 1989; CARAS et al. 1989; SMITH et al. 1990; MORAN et al. 1991), Qa-2 (WANECK et al. 1988b; ULKER et al. 1990), placental alkaline phosphatase (BERGER et al. 1988; MICANOVIC et al. 1990a, b), IgG F_c receptor type III (KUROSAKI and RAVETCH 1989; LANIER et al. 1989), and Ly-6E (SU and BOTHWELL 1989). Since DAF is one of the subjects of this volume and it has been analyzed extensively, we will present the data and conclusions concerning the signal for GPI anchoring in DAF, but the other work referenced supports these conclusions in general.

CARAS and colleagues have analyzed a series of DAF mutants to probe the nature of the consensus signal for GPI anchoring (CARAS and WEDDELL 1989; CARAS et al. 1989; MORAN et al. 1991); work from our laboratory has also addressed this question (SMITH et al. 1990; LUBLIN et al., unpublished observations). The first point is that the 37 COOH-terminal amino acids of human DAF fused to a secreted protein create a GPI-anchored protein (CARAS et al. 1987). These amino acids from DAF (shown in Table 1), comprising the 17 COOH-terminal hydrophobic amino acids and the preceding 20 amino acid hydrophilic domain, therefore contain the entire signal for GPI anchor attachment. The hydrophobic amino acid domain is necessary but not sufficient for anchoring (CARAS et al. 1989). As noted above, there is no specific sequence requirements for the hydrophobic domain as many other hydrophobic sequences can substitute for this part of the anchoring signal. However, it must have a minimum length; in the case of DAF, there is a sharp requirement for 13 of the 17 hydrophobic amino acids (Table 1) (SMITH et al. 1990).

Table 1. Analysis of two-part signal for GPI anchoring in DAF

Variant	COOH-terminal amino acid sequence		GPI anchor (%)
	Cleavage/attachment domain ↓	Hydrophobic domain	
DAF	PNKGSGTTSGTTRLLSGHTCF	TLTGLLGTLVTMGLLT	+(100)
DAF-13	--------------------	---FTLTGLLGTLVTM	+(92)
DAF-12	--------------------	----FTLTGLLGTLVT	+/−(11)
DAF-11	--------------------	-----FTLTGLLGTLV	+/−(6)
DAF-10	--------------------	------FTLTGLLGTL	−(2)
DAF-9	--------------------	-------FTLTGLLGT	−(<1)
DAF-MCP-18	--------------------	VWVIAVIVIAIVVGVAVI	+
hGH-DAF-37	PNKGSGTTSGTTRLLSGHTCFTLTGLLGTLVTMGLLT		+
hGH-DAF-35	-------------------------------------		+
hGH-DAF-33	---------------------------------		+
hGH-DAF-31	-------------------------------		+
hGH-DAF-29	-----------------------------		+
hGH-DAF-28	----------------------------		−
hGH-DAF-29-Gly	G----------------------------		+
hGH-DAF-29-Ala	A----------------------------		+
hGH-DAF-29-Asp	D----------------------------		+
hGH-DAF-29-Asn	N----------------------------		+
hGH-DAF-29-Group2	2----------------------------		+/−
hGH-DAF-29-Group3	3----------------------------		−

The 37 COOH-terminal amino acids of human DAF (shown here in standard single letter code) comprise a signal for GPI anchoring (CARAS et al. 1987) consisting of a cleavage/attachment domain and a hydrophobic domain (CARAS et al. 1989; SMITH et al. 1990; MORAN et al. 1991). The *arrow* shows the actual cleavage/attachment site in DAF (MORAN et al. 1991). The variant sequences below (*hyphens* represent wild-type sequence and blanks represent deletions) define some of the characteristics of each part of the signal, with the *upper set* being variants in the hydrophobic domain tested in the context of DAF (SMITH et al. 1990) and the *lower set* being variants in the cleavage/attachment domain tested in the context of a human growth hormone-DAF fusion construct (MORAN et al. 1991). The presence (+) or absence (−) of a GPI anchor on the variant is noted. For the variants in the hydrophobic region, the 17 hydrophobic amino acids are progressively shortened, and the actual efficiency of GPI anchoring compared to wild-type DAF is also given. The DAF-MCP construct replaces the entire hydrophobic domain of DAF with the first 18 amino acids of the MCP transmembrane hydrophobic domain. The cleavage/attachment site variants include group 1 amino acids (S, G, A, D, N), which support efficient GPI anchoring; group 2 amino acids (V, E, C, M), which only weakly support GPI anchoring; and group 3 amino acid (all 11 other amino acids) which do not support GPI anchoring

Since the hydrophobic domain by itself is not sufficient for GPI anchoring of DAF, there must be a second part to the consensus signal (CARAS et al. 1989; SMITH et al. 1990). Recent work has shown that the GPI attachment site in DAF is Ser-319 (MORAN et al. 1991) located in the hydrophilic amino acid region NH_2-terminal to the hydrophobic domain (Table 1), suggesting that the second part of the signal involves the cleavage/attachment domain. A detailed analysis of this cleavage/attachment site (MORAN et al. 1991) demonstrated that there is no apparent requirement for specific sequences NH_2-terminal to the attachment site but that, at the attachment site, there is a requirement for Ala, Asp, Asn, Gly, or the wild-type Ser for efficient GPI anchoring (Table 1); all other amino acids at this position lead to little or no GPI anchor attachment. This supports a suggestion that the small amino acids Ala, Asp, Asn, Cys, Gly, or Ser are necessary at the GPI attachment site, based on the observation that only these amino acids have been found at the attachment site in natural GPI-anchored proteins (CROSS 1990). Near saturation mutagenesis of the attachment site Asp of human placental alkaline phosphatase also found that only these six amino acids function well at this site (MICANOVIC et al. 1990a). Further analysis of the alkaline phosphatase mutants in a cell-free translation and processing system indicated that this specificity is enzymatically determined, i.e., reflects a requirement of the putative transmidase that cleaves the protein and attaches the GPI anchor (MICANOVIC et al. 1990b).

These data support a two-part consensus signal for GPI anchoring: (1) a COOH-terminal hydrophobic sequence of minimum length (13 amino acids for DAF) and (2) a cleavage/attachment domain requiring a small amino acid (Ala, Asp, Asn, Cys, Gly, or Ser) at the attachment site. There are probably further requirements of the amino acids around the cleavage/attachment site; again, inspection of the known GPI-anchored proteins suggests that similar small amino acids are usually found at the first and second positions of the cleaved peptide tail (CROSS 1990). Further mutagenesis experiments will be required to define the exact requirements for this small amino acid domain in the cleavage/attachment site (MICANOVIC et al. 1990b) and to define any additional constraints on the relation (distance) between the two parts of the signal. The full consensus signal for GPI anchor attachment probably depends on protein structural and physical properties, such as conformation and hydrophobicity, that are not completely apparent in the linear amino acid sequence, allowing a degree of latitude such as seen in NH_2-terminal signal sequences (VON HEIJNE 1986).

2.4 Paroxysmal Nocturnal Hemoglobinuria and Other Defects in the Pathway for GPI Anchoring

The pathway for synthesis of the GPI anchor and attachment to the protein is complex, involving many substrates and enzymes (not yet isolated) as described above. It is therefore not surprising that defects in the pathway have arisen naturally or can be created by mutagenesis. The net result of such a defect is an inability to express any of the GPI-anchored proteins on the cell surface. Paroxysmal nocturnal hemoglobinuria (PNH) is an acquired hemolytic anemia characterized by a population of blood cells that are abnormally sensitive to complement. The observation that two of the proteins that are lacking on PNH erythrocytes, AChE (AUDITORE et al. 1959; CHOW et al. 1985) and DAF (NICHOLSON-WELLER et al. 1983; PANGBURN et al. 1983), are both GPI-anchored proteins (HAAS et al. 1986; STIEGER et al. 1986; DAVITZ et al. 1986; MEDOF et al. 1986) suggested that the defect in PNH might involve a step in the GPI anchoring pathway (DAVITZ et al. 1986). Subsequent demonstration that multiple GPI-anchored proteins are absent from PNH cells confirms this hypothesis, but the actual defect (or heterogeneous defects) remains to be identified. (Further details on PNH can be found in the accompanying chapter in this volume.)

Several different cell lines that are deficient in GPI anchoring have been created by mutagenesis. The most widely studied are a series of murine lymphoma mutants isolated by Hyman and colleagues using mutagenesis followed by negative selection with anti-Thy-1 and complement (HYMAN 1973; HYMAN and STALLINGS 1974; HYMAN 1985). Complementation analysis using somatic cell hybrids between different mutants has divided these cell lines into eight complementation classes (A-H) (TROWBRIDGE et al. 1978; HYMAN 1985). Only class D has an apparent deletion of the Thy-1 structural gene, and the other classes presumably represent defects in the GPI anchoring pathway. Detailed biochemical analysis has shown that the class E mutant cells fail to synthesize dolichol-P-mannose, resulting in dolichol-linked oligosaccharides with five instead of nine mannose residues (CHAPMAN et al. 1979, 1980). Since dolichol-P-mannose is also the donor for the Man_3 core of the GPI anchor (MENON et al. 1990a), this explains the failure to express GPI-anchored proteins in the class E mutant cells. More recent biochemical studies of six of the complementation classes (A, B, C, E, F, and H) (FATEMI and TARTAKOFF 1986, 1988; CONZELMANN et al. 1986, 1988b) demonstrate that none of the other classes has the defect in dolichol-linked oligosaccharides seen in cells of class E. None of the cells of these five classes attach a full GPI anchor to Thy-1, although a partial anchor (without fatty acids and not actually able to anchor the cell to the membrane) is detected in class B and possibly class C.

Several other lymphocyte cell lines established by mutagenesis fail to express GPI-anchored membrane proteins (YEH et al. 1988; HOLLANDER et al. 1988). One of these, a murine T cell hybridoma mutant, failed to synthesize dolichol-P-mannose, and transfection with the yeast dolichol-P-mannose synthase gene corrects the synthesis of both dolichol-P-mannose and the GPI anchor, leading to surface expression of GPI-anchored proteins Thy-1 and Ly-6A (DEGASPERI et al. 1990). It has also been noted that the aminoglycoside G418 can correct the GPI anchoring defect in class B and F mutant cells (GUPTA et al. 1988), but the basis for this surprising finding is uncertain. A cloned natural killer cell line has been established from a PNH patient (SCHUBERT et al. 1990). These PNH and mutant cell lines should prove useful for further study of both GPI anchor biochemistry and analysis of the function of GPI anchors.

3 Functions of GPI Anchors

While great progress was being made on the structure of GPI-anchored membrane proteins, little could be said about the function of these proteins, in particular the role of the GPI anchor. The complexity of the GPI anchor (Fig. 1), the associated large number of enzymatic reactions needed for its synthesis and attachment to proteins, and the conservation of this structure through large evolutionary distances would argue for its necessity, but what are the functions it subserves? To be sure, the anchor serves as a mode of membrane attachment, but so does a transmembrane hydrophobic polypeptide sequence. It is important to put the question of function in perspective, realizing that the GPI anchor serves different functions on different proteins in addition to just attaching the protein to the plasma membrane. One must also separate the question of the function of a specific GPI-anchored protein (often not yet known) from the function of the GPI anchor, realizing that many GPI-anchored membrane proteins probably carry out functions for which the anchor is not important.

Once a possible functional role for the GPI anchor is proposed, the main approach to investigating this role is to compare versions of the same protein (ectodomain) attached to the cell through either a GPI or a transmembrane anchor, arguing that any functional difference observed must map to the different forms of membrane anchoring. These sets of proteins either arise as two naturally occurring versions of the protein or in most cases are produced through expression of cDNA constructs in either transfected cells or transgenic animals. This has produced data on several functional roles for the GPI anchor as will be discussed here. It should also be pointed out that this separate class of GPI-anchored membrane proteins might serve a more global role, in that these proteins do not have transmembrane and cytoplasmic domains to clutter the interior of the membrane and its intracytoplasmic face. For proteins that do not need to interact with other molecules in these regions, this prevents possible interference and thus might be a useful function from the point of view of the cell membrane as a whole (CROSS 1990).

3.1 Membrane Attachment and Release

Whatever else it might do, the GPI anchor does attach the protein to the cell membrane. This attachment yields an integral membrane protein; in general, only the conditions that release transmembrane proteins (e.g., detergents) will release GPI-anchored proteins. However, the use of a GPI anchor instead of a transmembrane polypeptide domain does allow differential protein release mechanisms. This has been considered for trypanosome VSG in which the genome encodes approximately 1000 alternate copies of VSG, allowing for evasion of host defenses by rapid replacement of one VSG with another. Since the trypanosome also encodes a PLC enzyme specific for GPI structures (BULOW and OVERATH 1986; FOX et al. 1986; HERELD et al. 1986), it was hypothesized that use of the GPI anchor in VSG together with this GPI-specific PLC provides a method to remove VSG rapidly from the cell surface during antigenic variation without any effect on transmembrane proteins. However, there is no direct evidence to date for this model. Several of the mammalian GPI-anchored membrane proteins also exist in a soluble form in plasma or other body fluids, such as is found for DAF (MEDOF et al. 1987). Again it is not known if the soluble protein arises from the membrane form by cleavage by mammalian glycan-PI-specific phospholipase C (Fox et al. 1987) or glycan-PI-specific phospholipase D (DAVITZ et al. 1987; Low and PRASAD 1988).

This proposed role of GPI anchors in the release of membrane proteins would serve as a mechanism to distinguish two classes of membrane proteins on the cell surface, GPI-anchored and transmembrane. A more subtle differentiation is also possible within the class of GPI-anchored membrane proteins. Variation in the fatty acids attached to glycerol would allow alterations in the overall hydrophobicity, potentially leading to changes in the strength of membrane attachment. Additionally, some GPI anchors possess an extra acyl group in ester linkage to a hydroxyl group of the inositol ring (Fig. 1) (ROBERTS et al. 1988a; MAYOR et al. 1990b). Not only does this lead to resistance to PI–PLC, raising the possibility of differential control of protein release from the membrane, but it would also produce a change in the hydrophobicity of membrane attachment groups as mentioned above. Thus, these changes in number and composition of fatty acids in the GPI anchor could be a mechanism for altering the amount of a GPI-anchored protein in the membrane through direct loss to the aqueous phase or removal by phospholipases. Cell-specific heterogeneity in sensitivity to phospholipase (Low et al. 1988) and changes occurring during cell differentiation or activation (PRESKY et al. 1990) have been observed, but it remains conjectural as to the physiological role of these variations.

3.2 Protein Lateral Mobility

Measurements of lateral diffusion of membrane glycoproteins fall in a range considerably slower than those observed for lipid probes (Table 2). Reconstitution of purified membrane proteins into artificial liposomes results in higher

Table 2. Lateral diffusion coefficients of membrane proteins and lipids

Molecule	$D \times 10^{10}$ cm^2/s	Reference
Lipid in artificial membrane	100–3000	MCCLOSKEY and POO 1986
Protein in artificial membrane	100–300	MCCLOSKEY and POO 1986
Lipid in plasma membrane	10–300	MCCLOSKEY and POO 1986
Protein in plasma membrane	0.1–50	MCCLOSKEY and POO 1986
GPI-anchored proteins		
Murine Thy-1	20–40	ISHIHARA et al. 1987
Rat alkaline phophatase	6–18	NODA et al. 1987
Human DAF	16	THOMAS et al. 1987
Trypanosome VSG	0.4–1	BULOW et al. 1988
VSG in hamster kidney cells	0.7	BULOW et al. 1988
Guinea pig sperm PH–20		
Testicular	0.2	PHELPS et al. 1988
Epididymal, pre-acrosome	1.8	COWAN et al. 1987
Epididymal, post-acrosome	49	COWAN et al. 1987

lateral diffusion coefficients, suggesting that there exist additional constraints on the lateral diffusion of proteins in plasma membranes of cells. When the lateral diffusion coefficients for several GPI-anchored membrane proteins were measured and found to be in the higher range typical of lipids, it was suggested that the GPI anchor confers an intrinsic high lateral mobility and that this was in fact one of the functions of this form of membrane anchoring. This fit in with the function of several of the GPI-anchored membrane proteins that are receptors or adhesion molecules and hence must move in the membrane in order to contact ligand. However, more recent work has cast doubt on this simple picture.

The measurements of lateral diffusion coefficients were all done by the technique of fluorescence recovery after photobleaching. Values $\geq 1 \times 10^{-9}$ for murine Thy-1, rat alkaline phosphatase, and human DAF (Table 2) were more typical of lipids than most previously measured proteins, suggesting that the GPI anchor conferred a high lateral mobility and might be a physiological reason to anchor a specific membrane protein in this manner. These measurements were done in a variety of cells expressing either endogenous or transfected protein. In addition, not all of the protein was in the mobile fraction exhibiting this high lateral diffusion. The lack of direct interactions between a GPI-anchored protein and the cytoskeleton or the transmembrane portions of other proteins was hypothesized to underlie the high lateral mobility. However, it must be noted that no direct comparisons were made between a GPI-anchored protein and the same extracellular domains anchored by a hydrophobic transmembrane polypeptide and cytoplasmic tail, so one could not conclude that the GPI anchor was the major or only factor in the high lateral mobility.

Measurements of lateral diffusion coefficients of trypanosome VSG gave lower values of $0.4 - 1.0 \times 10^{-10}$, with no significant difference between endogenous VSG in trypanosomes or VSG implanted into hamster kidney cells (Table 2) (BULOW et al. 1988). The sperm membrane protein PH-20, involved in sperm adhesion to the egg zona pellucida, is GPI-anchored and yet shows a very slow diffusion coefficient of 1.9×10^{-11} on testicular sperm (PHELPS et al. 1988). Interestingly, the lateral mobility increases markedly as the sperm matures, with a lateral diffusion coefficient of epididymal sperm prior to the acrosome reaction of 1.8×10^{-10} and 4.9×10^{-9} after the acrosome reaction (COWAN et al. 1987). Thus, not only can GPI-anchored proteins have lateral mobilities similar to transmembrane proteins, but the mobility can change due to some interaction involving the protein ectodomain. Work with the transmembrane class I antigen H-2L^d had shown that removal of almost all of the cytoplasmic tail did not significantly affect its lateral diffusion coefficient (EDIDIN and ZUNIGA 1984), yet removal of its three N-linked oligosaccharides caused a threefold increase in the lateral diffusion coefficient (WIER and EDIDIN 1988). Thus, both the recent work on GPI-anchored proteins and work on transmembrane proteins indicate that interactions that involve the protein ectodomain can be largely responsible for the lateral mobility of the protein. This does not rule out additional constraints on lateral mobility due to the transmembrane and cytoplasmic domains (absent from GPI-anchored proteins); for example, lateral mobility of erythrocyte band 3 is increased 100-fold in spectrin-deficient mouse erythrocytes (SHEETZ et al. 1980). For proteins that do not interact strongly with the cytoskeleton, however, the largest factor limiting mobility appears to be the interactions of the protein ectodomain with other molecules, independent of the mode of membrane attachment.

Direct comparison of GPI-anchored and transmembrane versions of the same protein are really required to address this issue of the role of the GPI anchor in protein lateral mobility. This has not yet been doen by direct biophysical measurements of lateral diffusion coefficients, but work from our laboratory investigated this issue with functional measurements (LUBLIN and COYNE 1991). DAF protects cells from complement-mediated damage by inhibiting the C3 convertase (NICHOLSON-WELLER et al. 1982; MEDOF et al. 1984). We found that transfection of DAF into Chinese hamster ovary (CHO) cells resulted in protection of the cells from cytotoxicity due to antibody plus human complement, with increasing protection from increasing amounts of DAF. A transmembrane version of DAF (constructed by replacing the COOH-terminal end of DAF with the transmembrane and cytoplasmic domains from either MCP or HLA-B44) was equivalent to GPI-anchored DAF in protection of CHO from complement-mediated cytotoxicity (LUBLIN and COYNE 1991). Similarly, creation of a GPI-anchored version of MCP (by using the COOH-terminal end of DAF that encodes the GPI attachment signal) allowed the demonstration that MCP protected cells from complement-mediated cytotoxicity equally well with either its natural transmembrane attachment or with a GPI anchor. Thus, DAF and MCP, which must move over the cell surface to contact C3b and C4b, show functional equivalence as either GPI-anchored or transmembrane proteins, suggesting equal lateral mobilities for either mode of membrane attachment.

3.3 Cell Activation

Even before it was established that Thy-1 possessed a GPI anchor, it was demonstrated that antibodies to Thy-1 could produce T cell activation (MAINO et al. 1981; GUNTER et al. 1984; MCDONALD et al. 1985). More than a dozen GPI-anchored membrane proteins have now been implicated in cell activation pathways, mostly in T cells but also in B cells, monocytes, and granulocytes (Table 3). The data generally involve the demonstration that polyclonal or monoclonal antibody to a T cell surface antigen can induce proliferation of peripheral T cells in the presence of a second cross-linking antibody (or accessory cells bearing F_c receptors) plus submitogenic amounts of a phorbol ester such as PMA. Two monoclonal antibodies directed to different epitopes would suffice without a second antibody for cross-linking. Besides proliferation, other indicators of cellular activation have been measured, including increases in intracellular calcium or inositol phosphates (KROCZEK et al. 1986; SEAMAN et al. 1991), production of cytokines or cytokine receptors such as IL-2 and IL-2 receptor (MALEK et al. 1986), triggering of the respiratory burst in granulocytes

Table 3. GPI-anchored proteins implicated in cell signaling/activation

Protein	Reference
Murine Thy-1	MAINO et al. 1981; GUNTER et al. 1984; MCDONALD et al. 1985
Murine Ly-6/TAP	ROCK et al. 1986; MALEK et al. 1986
Rat RT-6	WONIGKEIT and SCHWINZER 1987
Murine Qa-2	HAHN and SOLOSKI 1989; ROBINSON et al. 1989
Human DAF	DAVIS et al. 1988
Human CD14	MACINTYRE et al. 1989
Human CD73	THOMPSON et al. 1989
Human LFA-3	LE et al. 1987
Scrapie prion protein	CASHMAN et al. 1990
Human CD24	FISCHER et al. 1990
Human H19/MIRL	GROUX et al. 1989
Rat gp42	SEAMAN et al. 1991

(FISCHER et al. 1990), or inhibition of mitogenic response to lectins such as ConA (CASHMAN et al. 1990). A related finding is that activation by mitogens such as ConA and PHA is inhibited in mutant cell lines with defects in the GPI anchoring pathway (YEH et al. 1988) or by removal of GPI-anchored proteins from T cells by PI–PLC treatment (STIERNBERG et al. 1987; PRESKY et al. 1990).

The list of GPI-anchored proteins that can participate in cell activation (Table 3) let to the hypothesis that the GPI anchor was directly involved in signaling pathways. Direct evidence for this came from two groups that compared GPI-anchored and transmembrane versions of a protein, with only the GPI-anchored version able to signal (ROBINSON et al. 1989; SU et al. 1991). Studies on Qa-2, a GPI-anchored murine class I MHC antigen, showed that peripheral T cells from transgenic mice can be activated by antibody to Qa-2 but that a transmembrane version of Qa-2 could not be activated (ROBINSON et al. 1989). Cross-linking of a GPI-anchored version of H-2D^b, but not the wild-type transmembrane H-2D^b, could also drive cells to proliferate. Studies on the murine antigen Ly-6E transfected into a T cell helper clone demonstrated that the GPI-anchored Ly-6E, but not a transmembrane version (anchored by domains from H-2D^b), could mediate T cell activation as evidenced by both cell proliferation and IL-4 secretion (SU et al. 1991).

Although these experiments directly demonstrate that the GPI anchor is critical for this pathway of T cell activation, they do not establish the mechanism of signaling. Triggering of T cells by antigen proceeds through the T cell receptor (TCR)/CD3 complex that crosses the cell membrane and can generate intracellular signals. For this alternate pathway of T cell activation through GPI-anchored proteins, the immediate problem is how a molecule that does not cross the bilayer can generate an intracellular signal. There is no direct data to address this question at present, but two general models can be considered. Since the activation requires cross-linking and hence clustering of molecules on the cell surface, this might lead to internalization and/or release of the protein through action of a phospholipase, PI–PLC or PI–PLD. Data on Ly-6 (also designated T cell activating protein or TAP) (BAMEZAI et al. 1989) and CD73 (MASSAIA et al. 1990) indicate that soluble, but not immobilized, antibody can activate T cells, suggesting the possible requirement for internalization. However, immobilized but not soluble antibody to CD73 can act synergistically with signals generated through CD3 or CD2 (MASSAIA et al. 1990). Action of a PI–PLC or PI–PLD on the GPI-anchored proteins, either on the surface or after internalization, could generate diacylglycerol (an activator of protein kinase C) or phosphatidic acid, respectively, both of which might function as second messengers in cell activation. As an alternate model, the signaling might proceed through an associated transducer molecule. Since it appears that any GPI-anchored protein can use this signaling pathway, the transducer would have to recognize a common structure in these molecules, i.e., the GPI anchor. There is no direct evidence for such a transducer. Nonetheless, studies in TCR/CD3-negative mutants implicate the requirement for an intact TCR/CD3 complex in signaling through GPI-anchored proteins (GUNTER et al. 1987; SUSSMAN et al. 1988; BAMEZAI et al. 1988), suggesting that there may be a common final pathway for T cell activation initiated by antigen or through GPI-anchored proteins. Clearly much work remains to elucidate this latter pathway of T cell activation. As a separate but related point, it is of interest to note that a glycolipid with structural similarities to the GPI anchor, but not attached to protein, has been implicated as an intracellular second messenger for insulin (SALTIEL and CUATRECASAS 1986; SALTIEL et al. 1986).

3.4 Sorting in Polarized Epithelial Cells

Polarized epithelial cells form tight junctions that provide a barrier and permit vectorial movement of ions and macromolecules. The directional polarization defines an apical surface that faces the external environment and an inward facing basolateral surface. A major question is the nature of the signals that direct sorting in epithelial cells (reviewed in SIMONS and WANDINGER-NESS 1990). The work to be reviewed here indicates that the GPI anchor can act as an apical sorting signal; since non-GPI-anchored proteins can also have an apical distribution, this is not the only apical sorting signal.

The first indication that the GPI anchor itself acts as an apical sorting signal came from observations in the Madin-Darby canine kidney (MDCK) cell line. This cell line is well suited for studies of polarized epithelium because it will form monolayers with tight junctions when grown on filters, permitting differential labeling of the apical and basolateral surfaces. All of the endogenous GPI-anchored proteins detected in the MDCK cell line were restricted to the apical surface (LISANTI et al. 1988). Molecular biological techniques were then used to show that the GPI anchor itself conferred this apical expression on proteins expressed by transfection into MDCK cells (BROWN et al. 1989; LISANTI et al. 1989). Specifically, vesicular stomatitis virus glycoprotein G is normally expressed basolaterally, but conversion to a GPI-anchored protein (using the COOH-terminal sequence of Thy-1 as the GPI signal sequence) led to an apical distribution (BROWN et al. 1989). Conversely, expression of the GPI-anchored protein placental alkaline phosphatase as a transmembrane protein changed it from an apical to a basolateral protein (BROWN et al. 1989). Similarly, both DAF and GPI-anchored versions (created with the COOH-terminal sequence of DAF) of the basolateral protein herpes simplex glycoprotein D or the secreted protein human growth hormone are expressed as apical proteins (LISANTI et al. 1989). Further study of this GPI-anchored version of glycoprotein D in MDCK cells demonstrated that apical expression arose from apical delivery of the protein (as opposed to delivery to both membranes followed by preferential removal from the basolateral membrane or delivery to the basolateral membrane and subsequent transcytosis to the apical surface) (LISANTI et al. 1990). Glycosphingolipids are also expressed preferentially in the apical membrane (VAN MEER et al. 1987), raising the possibility that a similar sorting mechanism could direct both this lipid as well as the lipid-anchored proteins.

All of the above work was done in the MDCK cell line. Although this is a good model of polarized epithelium, it is not same as epithelial tissue in vivo. Only a few observations have been made in animals. Thy-1, when expressed in transgenic mice, had an apical distribution in renal epithelium (KOLLIAS et al. 1987). Although human DAF has an apical distribution when endogenously expressed in intestinal epithelial cell lines (LISANTI et al. 1989), when we expressed a transmembrane version of DAF (anchored by the transmembrane and cytoplasmic domains of HLA-B44) in transgenic mice, it had a uniform distribution over the apical and basolateral surfaces of intestinal epithelial cells (HANSBROUGH et al. 1991). Thus, the limited data from intact tissues is consistent with the more detailed studies in epithelial cell lines.

4 Conclusions

The past few years have seen not only a rapid growth in the number of known GPI-anchored membrane proteins but a wealth of new data on the structure and function of the GPI anchor. The elucidation of the complete structure of the GPI anchor was a biochemical tour de force. The conservation of the backbone of this structure across the large evolutionary gap from protozoa to humans highlights the importance of this form of membrane attachment. The biochemical pathway for synthesis of this anchor is being solved, and future work should yield the isolation, characterization, and eventual cloning of the enzymes involved. Some of the cells with apparent defects in this pathway, arising either in PNH cells or in mutant cell lines derived in tissue culture, should aid in this work. The nature of the consensus signal sequence at the COOH-terminal of the protein that directs GPI anchoring is also yielding to intense investigation, which has already delineated a two-part signal consisting of a hydrophobic segment (without a cytoplasmic tail in general) and a cleavage/attachment segment with certain preferred amino acids at the attachment site.

Corresponding progress on the functional roles of the GPI anchor has been much slower than this structural work, but it has now begun to produce direct results. Of course, the GPI anchor serves the functional role of membrane attachment, but other roles were speculative. Now, direct evidence obtained by comparing GPI-anchored and transmembrane versions of a protein has implicated the anchor in both lymphocyte activation and sorting in polarized epithelial cells. The long hypothesized role of the GPI anchor in increased lateral mobility of the protein remains in doubt and will require further analysis given the data reviewed here. These results form only the first step. With some definite functions for the GPI anchor documented, we must proceed to the even harder work of establishing the mechanisms whereby the GPI anchor mediates these functions. Given the ever increasing number of GPI-anchored proteins being identified, the results should justify the hard work.

Acknowledgements. The expert assistance of Jenny Boedeker with manuscript preparation is appreciated. Work from the author's laboratory was funded by National Institutes of Health grant GM41297 and by an Arthritis Investigator Award from the Arthritis Foundation.

References

Auditore JV, Hartmann RC (1959) Paroxysmal nocturnal hemoglobinuria: II. Erythrocyte acetylcholinesterase defect. Am J Med 27: 401–410
Bamezai A, Reiser H, Rock KL (1988) T cell receptor/CD3 negative variants are unresponsive to stimulation through the Ly-6 encoded molecule, TAP. J Immunol 141: 1423–1428
Bamezai A, Goldmacher V, Reiser H, Rock KL (1989) Internalization of phosphatidylinositol-anchored lymphocyte proteins. J Immunol 143: 3107–3116
Bangs JA, Hereld D, Krakow JL, Hart GW, Englund PT (1985) Rapid processing of the carboxyl terminus of a trypanosome variant surface glycoprotein. Proc Natl Acad Sci USA 82: 3207–3211
Bangs JD, Doering TL, Englund PT, Hart GW (1988) Biosynthesis of a variant surface glycoprotein of *Trypansoma brucei*: processing of the glycolipid membrane anchor and N-linked oligosaccharides. J Biol Chem 263: 17697–17705
Berger J, Howard AD, Brink L, Gerber L, Haubert J et al. (1988) COOH-terminal requirements for the correct processing of a phosphatidylinositol glycan-anchored membrane protein. J Biol Chem 263: 10016–10021
Berger J, Micanovic R, Greenspan RJ, Udenfriend S (1989) Conversion of placental alkaline phosphatase from a phosphatidylinositol glycan-anchored protein to an integral membrane protein. Proc Natl Acad Sci USA 86: 1457–1460
Boothroyd JC, Paynter CA, Cross GAM, Bernards A, Borst P (1981) Variant surface glycoproteins of *Trypanosoma brucei* are synthesised with cleavable hydrophobic sequences at the carboxy and amino termini. Nucleic Acids Res 9: 4735–4743
Brown D, Crise B, Rose JK (1989) Mechanism of membrane anchoring affects polarized expression of two proteins in MDCK cells. Science 245: 1499–1501
Bulow R, Overath P (1986) Purification and characterization of the membrane-form variant surface glycoprotein hydrolase of *Trypanosoma brucei.* J Biol Chem 261: 11918–11923
Bulow R, Overath P, Davoust J (1988) Rapid lateral diffusion of the variant surface glycoprotein in the coat of *Trypanosoma brucei.* Biochemistry 27: 2384–2388
Caras IW, Weddell GN (1989) Signal peptide for protein secretion directing glycophospholipid membrane anchor attachment. Science 243: 1196–1198

Caras IW, Weddell GN, Davitz MA, Nussenzweig V, Martin DW (1987) Signal for attachment of a phospholipid membrane anchor in decay-accelerating factor. Science 238: 1280–1283

Caras IW, Weddell GN, Williams SR (1989) Analysis of the signal for attachment of a glycophospholipid membrane anchor. J Cell Biol 108: 1387–1396

Cashman NR, Loertscher R, Nalbantoglu J, Shaw I, Kascsak RJ, Bolton DC, Bendheim PE (1990) Cellular isoform of the scrapie agent protein participates in lymphocyte activation. Cell 61: 185–192

Chapman A, Trowbridge IS, Hyman R, Kornfeld S (1979) Structure of the lipid-linked oligosaccharides that accumulate in class E Thy-1 negative mutant lymphomas. Cell 17: 509–515

Chapman A, Fujimoto K, Kornfeld S (1980) The primary glycosylation defect in class E Thy-1-negative mutant mouse lymphoma cells is an inability to synthesize dolichol-P-mannose. J Biol Chem 255: 4441–4446

Chow FL, Telen MJ, Rosse WF (1985) The acetylcholinesterase defect in paroxysmal nocturnal hemoglobinuria: evidence that the enzyme is absent from the cell membrane. Blood 66: 940–945

Conzelmann A, Spiazzi A, Hyman R, Bron C (1986) Anchoring of membrane proteins via phosphatidylinositol is deficient in two classes of Thy-1 negative mutant lymphoma cells. EMBO J 5: 3291–3296

Conzelmann A, Riezman H, Desponds C, Bron C (1988a) A major 125-kd membrane glycoprotein of *Saccharomyces cerevisiae* is attached to the lipid bilayer through an inositol-containing phospholipid. EMBO J 7: 2233–2240

Conzelmann A, Spiazzi A, Bron C, Hyman R (1988b) No glycolipid anchors are added to Thy-1 glycoprotein in Thy-1 negative mutant thymoma cells of four different complementation classes. Mol Cell Biol 8: 674–678

Cowan AE, Myles DG, Koppel DE (1987) Lateral diffusion of the PH-20 protein on guinea pig sperm: evidence that barriers to diffusion maintain plasma membrane domains in mammalian sperm. J Cell Biol 104: 917–923

Cross GAM (1990) Glycolipid anchoring of plasma membrane proteins. Annu Rev Cell Biol 6: 1–39

Davis LS, Patel SS, Atkinson JP, Lipsky PE (1988) Decay-accelerating factor functions as a signal transducing molecule for human T cells. J Immunol 141: 2246–2252

Davitz MA, Low MG, Nussenzweig V (1986) Release of decay-accelerating factor (DAF) from the cell membrane by phosphatidylinositol-specific phospholipase C (PIPLC). Selective modification of a complement regulatory protein. J Exp Med 163: 1150–1161

Davitz MA, Hereld D, Shak S, Krakow J, Englund PT, Nussenzweig V (1987) A glycan-phosphatidylinositol-specific phospholipase D in human serum. Science 238: 81–84

DeGasperi R, Thomas LJ, Sugiyama E, Chang HM, Beck PJ, Orlean P, Albright C, Weneck G, Sambrook JF, Warren CD, Yeh ETH (1990) Correction of a defect in mammalian GPI anchor biosynthesis by a transfected yeast gene. Science 250: 988–991

Doering TL, Masterson WJ, Englund PT, Hart GW (1989) Biosynthesis of the glycosyl phosphatidylinositol membrane anchor of the trypanosome variant surface glycoprotein: origin of the nonacetylated glucosamine. J Biol Chem 264: 11168–11173

Edidin M, Zuniga M (1984) Lateral diffusion of wild-type and mutant L^d antigens in L cells. J Cell Biol 99: 2333–2335

Fasel N, Rousseaux M, Schaerer E, Medof ME, Tykocinski ML, Bron C (1989) In vitro attachment of glycosylinositolphospholipid anchor structures to mouse Thy-1 antigen and human decay-accelerating factor. Proc Natl Acad Sci USA 86: 6858–6862

Fatemi SH, Tartakoff AM (1986) Hydrophilic anchor-deficient Thy-1 is secreted by a class E mutant T lymphoma. Cell 46: 653–657

Fatemi SH, Tartakoff AM (1988) The phenotype of five classes of T lymphoma mutants: defective glycophospholipid anchoring, rapid degradation, and secretion of Thy-1 glycoprotein. J Biol Chem 263: 1288–1294

Ferguson MAJ, Williams AF (1988) Cell-surface anchoring of proteins via glycosyl-phosphatidylinositol structures. Annu Rev Biochem 57: 285–320

Ferguson MAJ, Haldar K, Cross GAM (1985) *Trypanosoma brucei* variant surface glycoprotein has a sn-1, 2-dimyristyl glycerol membrane anchor at its COOH-terminus. J Biol Chem 260: 4963–4968

Ferguson MAJ, Duszenko M, Lamont GS, Overath P, Cross GAM (1986) Biosynthesis of *Trypanosoma brucei* variant surface glycoproteins: N-glycosylation and addition of a phosphatidylinositol membrane anchor. J Biol Chem 261: 356–362

Ferguson MAJ, Homans SW, Dwek RA, Rademacher TW (1988) Glycosylphosphatidylinositol moiety that anchors *Trypanosoma brucei* variant surface glycoprotein to the membrane. Science 239: 753–759

Fischer GF, Majdic O, Gadd S, Knapp W (1990) Signal transduction in lymphocytic and myeloid cells via CD24, a new member of phosphoinositol-anchored membrane molecules. J Immunol 144: 638–641

Fox JA, Duszenko M, Ferguson MAJ, Low MG, Cross GAM (1986) Purification and characterization of a novel glycan-phosphatidylinositol-specific phospholipase C from *Trypanosoma brucei*. J Biol Chem 261: 15767–15771

Fox JA, Soliz NM, Saltiel AR (1987) Purification and characterization of a phosphatidylinositol glycan specific phospholipase C from hepatic plasma membranes. Proc Natl Acad Sci USA 84: 2663–2667

Groux H, Huet S, Aubrit F, Tran HC, Boumsell L, Bernard A (1989) A 19-kDa human erythrocyte molecule H19 is involved in rosettes, present on nucleated cells, and required for T cell activation: comparison of the roles of H19 and LFA-3 molecules in T cell activation. J Immunol 142: 3013–3020

Gunter KC, Malek TR, Shevach EM (1984) T cell-activating properties of an anti-Thy-1 monoclonal antibody. Possible analogy to OKT3/LEU-4. J Exp Med 159: 716–730

Gunter KC, Germain RN, Kroczek RA, Saito T, Yokoyama WM, Chan C, Weiss A, Shevach EM (1987) Thy-1-mediated T-cell activation requires co-expression of CD3/Ti complex. Nature 326: 505–507

Gupta D, Tartakoff A, Tisdale E (1988) Metabolic correction of defects in the lipid anchoring of Thy-1 in lymphoma mutants. Science 242: 1446–1448

Haas R, Brandt PT, Knight J, Rosenberry TL (1986) Identification of amine components in a glycolipid membrane-binding at the C-terminus of human erythrocyte acetylcholinesterase. Biochemistry 26: 3098–3105

Hahn AB, Soloski MJ (1989) Anti-Qa-2-induced T cell activation: the parameters of activation, the definition of mitogenic and nonmitogenic antibodies, and the differential effects on $CD4^+$ vs $CD8^+$ T cells. J Immunol 143: 407–413

Hansbrough JR, Lublin DM, Roth KA, Birkenmeier EA, Gordon JI (1991) A model for studying regional differentiation and protein sorting in the intestinal epithelium: expression of a liver fatty acid binding protein/human decay accelerating factor/HLA-B44 chimeric gene in transgenic mice. Am J Physiol 260: G929–G939

Hereld D, Krakow JL, Bangs JD, Hart GW, Englund PT (1986) A phospholipase C from *Trypanosoma brucei* which selectively cleaves the glycolipid on the variant surface glycoprotein. J Biol Chem 261: 13813–13819

Hollander N, Selvaraj P, Springor TA (1988) Biosynthesis and function of a LFA-3 in human mutant cells deficient in phosphatidylinositol-anchored proteins. J Immunol 141: 4283–4290

Homans SW, Ferguson MAJ, Dwek RA, Rademacher TW, Anand R, Williams AF (1988) Complete structure of the glycosyl phosphatidylinositol membrane anchor of rat brain Thy-1 glycoprotein. Nature 333: 269–272

Hyman R (1973) Studies on surface antigen variants. Isolation of the complementary variants for Thy-1.2. JNCI 50: 415–422

Hyman R (1985) Cell-surface antigen mutants of haematopoietic cells: tools to study differentiation, biosynthesis and function. Biochem J 225: 27–40

Hyman R, Stallings V (1974) Complementation patterns of Thy-1 variants and evidence that antigen-loss variants "pre-exist" in the parental population. JNCI 52: 429–436

Ishihara A, Hou Y, Jacobson K (1987) The Thy-1 antigen exhibits rapid lateral diffusion in the plasma membrane of rodent lymphoid cells and fibroblasts. Proc Natl Acad Sci USA 84: 1290–1293

Kollias D, Evans DJ, Ritter M, Beech J, Grosveld F (1987) Ectopic expression of Thy-1 in the kidneys of transgenic mice induces functional and proliferative abnormalities. Cell 51: 21–31

Krakow JL, Hereld D, Bangs JD, Hart GW, Englund PT (1986) Identification of a glycolipid precursor of the *Trypanosoma brucei* variant surface glycoprotein. J Biol Chem 261: 12147–12153

Kroczek RA, Gunter KC, Germain RN, Shevach EM (1986) Thy-1 functions as a signal transduction molecule in T lymphocytes and transfected B lymphocytes. Nature 322: 181–184

Kurosaki T, Ravetch JV (1989) A single amino acid in the glycosyl phosphatidylinositol attachment domain determines the membrane topology of FcγRIII. Nature 342: 805–807

Lanier LL, Cwirla S, Yu G, Testi R, Phillips JH (1989) Membrane anchoring of a human IgG Fc receptor (CD16) determined by a single amino acid. Science 246: 1611–1614

Le P, Denning S, Springer T, Haynes B, Singer K (1987) Anti-LFA-3 monoclonal antibody induces interleukin 1 (IL 1) release by thymic epithelial (TE) cells and monocytes. Fed Proc 46: 447

Lisanti MP, Sargiacomo MP, Graeve L, Saltiel AR, Rodriguez-Boulan E (1988) Polarized apical distribution of glycosylphosphatidylinositol-anchored proteins in a renal epithelial cell line. Proc Natl Acad Sci USA 85: 9557–9561

Lisanti MP, Caras IW, Davitz MA, Rodriguez-Boulan E (1989) A glycophospholipid membrane anchor acts as an apical targeting signal in polarized epithelial cells. J Cell Biol 109: 2145–2156

Lisanti MP, Caras IW, Gilbert T, Hanzel D, Rodriguez-Boulan E (1990) Vectorial apical delivery and slow endocytosis of a glycolipid-anchored fusion protein in transfected MDCK cells. Proc Natl Acad Sci USA 87: 7419–7423

Low MG (1987) Biochemistry of the glycosyl-phosphatidylinositol membrane protein anchors. Biochem J 244: 1–13

Low MG (1989) Glycosyl-phosphatidylinositol: a versatile anchor for cell surface proteins. FASEB J 3: 1600–1608

Low MG, Prasad ARS (1988) A phospholipase D specific for the phosphatidylinositol anchor of cell-surface proteins is abundant in plasma. Proc Natl Acad Sci USA 85: 980–984

Low MG, Saltiel AR (1988) Structural and functional roles of glycosyl phosphatidylinositol in membranes. Science 239: 268–275

Low MG, Stiernberg J, Waneck GL, Flavell RA, Kincade PW (1988) Cell-specific heterogeneity in sensitivity of phosphatidylinositol-anchored membrane antigens to release by phospholipase C. J Immunol Methods 113: 101–111

Lublin DM, Coyne KE (1991) Phospholipid-anchored and transmembrane versions of either decay accelerating factor or membrane cofactor protein show equal efficiency in protection from complement-mediated cell damage. J Exp Med 174: 35–44

MacIntyre EA, Roberts PJ, Jones M, van der Schoot CE, Favalaro EJ, Tidman N, Linch DC (1989) Activation of human monocytes occurs on cross-linking monocytic antigens to an Fc receptor. J Immunol 142: 2377–2383

Maino VC, Norcross MA, Perkins MS, Smith RT (1981) Mechanism of Thy-1-mediated T cell activation: roles of Fc Receptors, T200, Ia, and H-2 glycoproteins in accessory cell function. J Immunol 126: 1829–1836

Malek TR, Ortega G, Chan C, Kroczek RA, Shevach EM (1986) Role of Ly-6 in lymphocyte activation: II. Induction of T cell activation by monoclonal anti-Ly-6 antibodies. J Exp Med 164: 709–722

Massaia M, Perrin L, Bianchi A, Ruedi J, Attisano C, Altieri D, Rijkers GT, Thompson LF (1990) Human T cell activation: synergy between CD73 (ecto-5'-nucleotidase) and signals delivered through CD3 and CD2 molecules. J Immunol 145: 1664–1674

Masterson W, Doering TL, Hart GW, Englund PT (1989) A novel pathway for glycan assembly: biosynthesis of the glycosylphosphatidylinositol anchor of the trypanosome variant surface glycoprotein. Cell 56: 793–800

Masterson WJ, Raper J, Doering TL, Hart GW, Englund PT (1990) Fatty acid remodeling: a novel reaction sequence in the biosynthesis of trypanosome glycosyl phosphatidylinositol membrane anchors. Cell 62: 73–80

Mayor S, Menon AK, Cross GAM, Ferguson MAJ, Dwek RA, Rademacher TW (1990a) Glycolipid precursors for the membrane anchor of *Trypanosoma brucei* variant surface glycoproteins: I. Glycan structure of the phosphatidylinositol-specific phospholipase C sensitive and resistant glycolipids. J Biol Chem 265: 6164–6173

Mayor S, Menon AK, Cross GAM (1990b) Glcolipid precursors for the membrane anchor of *Trypanosoma brucei* variant surface glycoproteins: II. Lipid structures of the phosphatidylinositol-specific phospholipase C sensitive and resistant glycolipids. J Biol Chem 265: 6174–6181

McCloskey MA, Poo MM (1986) Rates of membrane-associated reactions: reduction of dimensionality revisited. J Cell Biol 102: 88–96

McDonald HR, Bron C, Rousseaux M, Horvath C, Cerottini JC (1985) Production and characterization of monoclonal anti-Thy-1 antibodies that stimulate lymphokine production by cytolytic T cell clones. Eur J Immunol 15: 495–501

Medof ME, Kinoshita T, Nussenzweig V (1984) Inhibition of complement activation on the surface of cells after incorporation of decay-accelerating factor (DAF) into their membranes. J Exp Med 160: 1558–1578

Medof ME, Walter EI, Roberts WL, Haas R, Rosenberry TL (1986) Decay accelerating factor of complement is anchored to cells by a C-terminal glycolipid. Biochemistry 25: 6740–6747

Medof ME, Walter EI, Rutgers JL, Knowles DM, Nussenzweig V (1987) Identification of the complement decay-accelerating factor (DAF) on epithelium and glandular cells and in body fluids. J Exp Med 165: 848

Menon AK, Mayor S, Ferguson MAJ, Duszenko M, Cross GAM (1988) Candidate glycophospholipid precursor for the glycophosphatidylinositol membrane anchor of *Trypanosoma brucei* variant surface glycoproteins. J Biol Chem 263: 1970–1977

Menon AK, Mayor S, Schwarz RT (1990a) Biosynthesis of glycosylphosphatidylinositol lipids in *Trypanosoma brucei*: involvement of mannosylphosphoryldolichol as the mannose donor. EMBO J 9: 4249–4258

Menon AK, Schwartz RT, Major S, Cross GAM (1990b) Cell-free synthesis of glycosylphosphatidylinositol precursors for the glycolipid membrane anchor of *Trypanosoma brucei* variant surface glycoproteins: structural charactrization of putative biosynthetic intermediates. J Biol Chem 265: 9033–9042

Micanovic R, Gerber LD, Berger J, Kodukula, K, Udenfriend S (1990a) Selectivity of the cleavage/attachment site of phospatidylinositol-glycan-anchored membrane proteins determined by site-specific mutagenesis at Asp-484 of placental alkaline phosphatase. Proc Natl Acad Sci USA 87: 1–6

Micanovic R, Kodukula K, Gerber LD, Udenfriend S (1990b) Selectivity at the cleavage/attachment site of phosphatidylinositolglycan anchored membrane proteins is enzymatically determined. Proc Natl Acad Sci USA 87: 7939–7943

Moran P, Raab H, Kohr WJ, Caras IW (1991) Glycophospholipid membrane anchor attachment: molecular analysis of the cleavage/attachment site. J Biol Chem 266: 1250–1257

Nicholson-Weller A, Burge J, Fearson DT, Weller PF, Austen KF (1982) Isolation of a human erythrocyte membrane glycoprotein with decay-accelerating activity for C3 convertases of the complement system. J. Immunol 129: 184

Nicholson-Weller A, March JP, Rosenfeld SI, Austen KF (1983) Affected erythrocytes of patients with paroxysmal nocturnal hemoglobinuria are deficient in the complement regulatory protein decay accelerating factor. Proc Natl Acad Sci USA 80: 5066–5070

Noda M, Yoon K, Rodan GA, Koppel DE (1987) High lateral mobility of endogenous and transfected alkaline phosphatase: a phosphatidylinositol-anchored membrane protein. J Cell Biol 105: 1671–1677

Ogata S, Hayashi Y, Takami N, Ikehara Y (1988) Chemical characterization of the membrane-anchoring domain of human placental alkaline phosphatase. J Biol Chem 263: 10489–10494

Pangburn MK, Schreiber RD, Muller-Eberhard HJ (1983) Deficiency of an erythrocyte membrane protein with complement regulator activity in paroxysmal nocturnal hemoglobinuria. Proc Natl Acad Sci USA 80: 5430–5434

Phelps BM, Primakoff P, Koppel DE, Low MG, Myles DG (1988) Restricted lateral diffusion of PH-20, a PI-anchored sperm membrane protein. Science 240: 1780–1782

Presky DH, Low MG, Shevach EM (1990) Role of phosphatidylinositol-anchored proteins in T cell activation. J Immunol 144: 860–868

Roberts WL, Kim BH, Rosenberry TL (1987) Differences in the glycolipid membrane anchors of bovine and human erythrocyte acetylcholinesterases. Proc Natl Acad Sci USA 84: 7817–7821

Roberts WL, Myher JJ, Kuksis A, Low MG, Rosenberry TL (1988a) Lipid analysis of the glycoinositol phospholipid membrane anchor of human erythrocyte acetylcholinesterase: palmitoylation of inositol results in resistance to phosphatidylinositol-specific phospholipase C. J Biol Chem 263: 18766–18775

Roberts WL, Santikarn S, Reinhold VN, Rosenberry TL (1988b) Structural characterization of the glycoinositol phospholipid membrane anchor of human erythrocyte acetylcholinesterase by fast atom bombardment mass spectrometry. J Biol Chem 263: 18776–18784

Robinson PJ, Millrain M, Antoniou J, Simpson E, Mellor AL (1989) A glycophospholipid anchor is required for Qa-2-mediated T cell activation. Nature 342: 85–87

Rock KL, Yeh ETH, Gramm CF, Haber SI, Reiser H, Benacerraf B (1986) TAP, a novel T cell-activating protein involved in the stimulation of MHC-restricted T lymphocytes. J Exp Med 163: 315–333

Saltiel AR, Cuatrecasas P (1986) Insulin stimulates the generation from hepatic plasma membranes of modulators derived from an inositol glycolipid. Proc Natl Acad Sci USA 83: 5793–5797

Saltiel AR, Fox JA, Sherline P, Cuatrecasas P (1986) Insulin-stimulated hydrolysis of a novel glycolipid generates modulators of cAMP phosphodiesterase. Science 233: 967–972

Schubert J, Uciechowski P, Delany P, Tischler HJ, Kolanus W, Schmidt RE (1990) The PIG-anchoring defect in NK lymphocytes of PNH patients. Blood 76: 1181–1187

Seaman WE, Niemi EC, Stark MR, Goldfien RD, Pollock AS, Imboden JB (1991) Molecular cloning of gp42, a cell-surface molecule that is selectively induced on rat natural killer cells by interleukin 2: glycolipid membrane anchoring and capacity for transmembrane signaling. J Exp Med 173: 251–260

Sheetz MP, Schindler M, Koppel DE (1980) Lateral mobility of integral membrane proteins is increased in spherocytic erythrocytes. Nature 285: 510–512

Simons K, Wandinger-Ness A (1990) Polarized sorting in epithelia. Cell 62: 207–210

Smith JD, Arce MA, Thompson ES, Lublin DM (1990) Analysis of the glycophospholipid (GPI) anchoring signal in decay accelerating factor (DAF). FASEB J 4: A2188

Stahl N, Borchelt DR, Hsiao K, Prusiner B (1987) Scrapie prion protein contains a phosphatidylinositol glycolipid. Cell 51: 229–240

Stieger A, Cardoso de Almeida ML, Blatter MC, Brodbeck U, Bordier C (1986) The membrane-anchoring systems of vertebrate acetylcholinesterase and variant surface glycoproteins of African trypanosomes share a common antigenic determinant. FEBS Lett 199: 182–187

Stiernberg J, Low MG, Flaherty L, Kincade PW (1987) Removal of lymphocyte surface molecules with phosphatidylinositol-specific phospholipase C: effects on mitogen response and evidence that ThB and certain Qa antigens are membrane-anchored via phosphatidylinositol. J Immunol 38: 3877–3884

Su B, Bothwell ALM (1989) Biosynthesis of a phosphatidylinositol-glycan-linked membrane protein: signals for post-translational processing of the Ly-6E antigen. Mol Cell Biol 9: 3369–3376

Su B, Waneck GL, Flavell RA, Bothwell ALM (1991) The glycosyl phosphatidylinositol anchor is critical for Ly-6A/E-mediated T cell activation. J Cell Biol 112: 377–384

Sussman JJ, Saito T, Shevach EM, Germain RN, Ashwell JD (1988) Thy-1- and Ly-6-mediated lymphokine production and growth inhibition of a T cell hybridoma require co-expression of the T cell antigen receptor complex. J Immunol 140: 2520–2526

Thomas J, Webb W, Davitz MA, Nussenzweig V (1987) Decay acclerating factor diffuses rapidly on HeLa cell surfaces. Biophys J 51: 522a

Thompson LF, Ruedi JM, Glass A, Low MG, Lucas AH (1989) Antibodies to 5'-nucleotidase (CD73), a glycosyl-phosphatidylinositol-anchored protein, cause human peripheral blood T cells to proliferate. J Immunol 143: 1815–1821

Trowbridge I, Hyman R, Mazauskas C (1978) The synthesis and properties of T25 glycoprotein in Thy-1-negative mutant lymphoma cells. Cell 14: 21–32

Tse AGD, Barclay AN, Watts A, Williams AF (1985) A glycophospholipid tail at the carboxyl terminus of the Thy-1 glycoprotein of neurons and thymocytes. Science 230: 1003–1008

Tykocinski ML, Shu HK, Ayers DJ, Walter EI, Getty RR et al. (1988) Glycolipid reanchoring of T-lymphocyte surface antigen CD8 using the 3' end sequence of decay-accelerating factor's mRNA. Proc Natl Acad Sci USA 85: 3555–3559

Ulker N, Hood LE, Stroynowski I (1990) Molecular signals for phosphatidylinositol modification of the Qa-2 antigen. J Immunol 145: 2214–2219

van Meer G, Stelzer EHK, Wijnaendts-van Resandt RW, Simons K (1987) Sorting of sphingolipids in epithelial (Madin-Darby canine kidney) cells. J Cell Biol 105: 1623–1635

von Heijne G (1986) A new method for predicting signal sequence cleavage sites. Nucleic Acids Res 14: 4683–4690

Walter EI, Roberts WL, Rosenberry TL, Ratnoff WD, Medof ME (1990) Structural basis for variations in the sensitivity of human decay-accelerating factor to phosphatidylinositol-specific phospholipase C cleavage. J Immunol 144: 1030–1036

Waneck GL, Sherman DH, Kincade PW, Low MG, Flavell RA (1988a) Molecular mapping of signals in the Qa-2 antigen required for attachment of the phosphatidylinositol membrane anchor. Proc Natl Acad Sci USA 85: 577–581

Waneck GL, Stein ME, Flavell RA (1988b) Conversion of a PI-anchored protein to an integral membrane protein by a single amino acid mutation. Science 241: 697–700

Wier M, Edidin M (1988) Constraint of the translational diffusion of a membrane glycoprotein by its external domains. Science 242: 412–414

Williams AF, Tse AGD, Gagnon J (1988) Squid glycoproteins with structural similarities to Thy-1 and Ly-6 antigens. Immunogenetics 27: 265–272

Wonigkeit K, Schwinzer R (1987) Induction of the transplantation response: polyclonal activation of rat T lymphocytes by RT6 alloantisera. Trans Proc 19: 296–299

Yeh, ETH, Reiser H, Bamezai A, Rock KL (1988) TAP transcription and phosphatidylinositol linkage mutants are defective in activation through the T cell receptor. Cell 52: 665–674

Paroxysmal Nocturnal Hemoglobinuria

W. F. Rosse

1 Introduction	163
2 Clinical Symptoms Related to the Defective Cells	164
3 Hemolysis and the Defect in Complement Regulation	165
4 Thrombosis	167
5 Increased Tendency of Infection	168
6 Deficient Hematopoiesis	169
7 Leukemia	169
8 Conclusions	170
References	170

1 Introduction

Paroxysmal nocturnal hemoglobinuria (PNH) is a clonal disorder of the hematopoietic stem cell (HARTMANN and ARNOLD 1977) resulting in the production of blood cells which are defective in that they lack or are markedly deficient in glycan-phosphatidylinositol (GPI)-linked surface proteins (ROSSE 1990a). To date, 11 such proteins have been found to be missing or deficient on the abnormal cells (AUDITORE et al. 1960; KUNSTLING and ROSSE 1969; BECK and VALENTINE 1951; BURROUGHS et al. 1988; NICHOLSON-WELLER et al. 1983a; PANGBURN et al. 1983; HOLGUIN et al. 1989a; HANSCH et al. 1987; SELVARAJ et al. 1987, 1988; SIMMONS et al. 1989; VAN DER SCHOOT et al. 1989) (see Table 1). Other GPI-linked proteins that have not been identified are probably also missing.

The disorder arises as an acquired somatic defect that affects only the blood cells, i.e., erythrocytes, platelets, granulocytes, monocytes, and lymphocytes (NICHOLSON-WELLER et al. 1983b; KINOSHITA et al. 1985). No evidence of genetic transmission of either the disease or a tendency for the disease has been reported. The proportion of the blood cells that are affected varies greatly from patient to patient and depends upon the degree to which the normal marrow precursors have been replaced by those that lead to the abnormal blood cells.

Division of Hematology/Oncology, Duke University School of Medicine, Durham, NC 27710, USA

Table 1. Glycan-phosphatidylinositol linked proteins deficient on PNH cells

Membrane enzymes
  Erythrocyte acetylcholinesterase
  Leukocyte alkaline phosphatase
  Lymphocyte 5'-ectonucleotidase
Complement regulatory proteins
  Decay Accelerating factor (CD55); protectin membrane inhibitor of reactive lysis, CD59)
  C8 binding protein (homologous restriction factor)
Immunologic proteins
  $Fc_{(\gamma)}$ receptor III
  Leukocyte function antigen 3
  Endotoxin binding protein receptor (CD14)
Miscellaneous
  CD24 (function unknown)
  CD48 (function unknown)

The nature of the defect is gradually becoming clear. Studies have shown that the mRNA for the deficient proteins is present in the abnormal PNH cells (STAFFORD et al. 1988) and that the variants of the proteins apparently lacking the GPI anchor are synthesized in the endoplasmic reticulum (CAROTHERS et al. 1990). More recent data has suggested that the defect resides in the ability to synthesize the GPI anchor; the granulocytes of patients with PNH are not able to make this moiety (MAHONEY et al. 1991). The exact step or steps in the biosynthetic sequence for GPI at which this defect occurs remains to be elucidated.

2 Clinical Symptoms Related to the Defective Cells

The clinical symptomatology of the disorder is remarkably diverse (DACIE and LEWIS 1972; SIRCHIA and LEWIS 1975; ROSSE 1990b). Most, if not all, patients with PNH exhibit hemolysis of the abnormal red cells. The bone marrow of most patients can be demonstrated to have less than normal hematopoietic potential; that is, the marrow is less able than normal to produce blood cells. Many patients have a tendency to thrombosis that results in venous thromboses in unusual sites. Some patients exhibit a minor but occasionally serious defect in the ability to fight infections. A very small proportion of patients ultimately have acute leukemia.

PNH was classified among the myeloproliferative syndromes (chronic myelogenous leukemia, essential thrombocythemia, polycythemia rubra vera) DAMESHEK 1969), but these stem cell disorders are characterized by overproduction of one or more cellular elements rather than underproduction as is often the case in PNH. More recently, PNH has been classified among the myelodysplastic syndromes (ROSSE 1980). These syndromes, which include myelofibrosis, refractory anemia, sideroblastic anemia, and erythroleukemia, are all stem cell disorders characterized by cytopenia due to diminished effective hematopoiesis, the production of abnormal blood cells in all three myeloid cell lines, and the eventual evolution into acute leukemia in some patients. PNH may rarely precede or follow any of these stem cell disorders.

Here, the clinical findings in PNH will be described in some detail and related, when possible, to the known defects of membrane proteins. Based on our current state of knowledge, success in doing so will more or less depend upon the symptom and the defect.

3 Hemolysis and the Defect in Complement Regulation

The characteristic that has been used to identify PNH has been the unusual sensitivity of red cells to lysis by activated complement (ROSSE and PARKER 1986). This was first observed in 1913 by HIJMANS VAN DEN BERGH (1911), although he was not able to prove that lysis of the patient's red cells in acidified serum was due to activation of serum complement. HAM and his associates demonstrated the role of complement activation and prescribed the test most commonly used to diagnose the disorder—the acidified serum lysis test—in which complement is activated by lowering the pH of serum (HAM 1937; HAM and DINGLE 1939). This test, rendered more sensitive by optimizing the concentration of magnesium (MAY et al. 1973), has remained the standard diagnostic test for the disorder.

The peculiar sensitivity to complement lysis was more quantitatively demonstrated in the complement lysis sensitivity test (CLS) of ROSSE and DACIE (1966), in which graded amounts of complement were activated by a complement-fixing antibody. These studies initially demonstrated that the blood of patients with PNH contained cells that were normal or nearly normal in their sensitivity to complement (PNH I cells) and cells that were markedly more sensitive to complement (PNH III cells). Cells of intermediate sensitivity were later identified and called PNH II (ROSSE et al. 1974a) or PNH IIIa (CHOW et al. 1986) cells. It is now recognized that a full spectrum of cells of intermediate sensitivity, ranging from that of PNH I to that of PNH III cells, may occur in some patients (ROSSE et al. 1991).

The sensitivity to complement arises from the lack of two (possibly three) GPI-linked proteins from the surface of the cells: (1) decay accelerating factor (DAF; CD55) (NICHOLSON-WELLER et al. 1983a; PANGBURN et al. 1983); (2) protectin (membrane inhibitor of reactive lysis, MIRL; CD59) (HOLGUIN et al. 1989a); and (3) C8 binding protein (C8-bp; homologous restriction factor) (HÄNSCH et al. 1987). The identity and characteristics of the first two are well established; those of the third are less so.

The markedly sensitive PNH III red cells lack all evidence of DAF and MIRL, either by flow cytometry or by membrane extraction (ROSSE et al. 1991). This results in: (a) markedly enhanced activation of the amplification steps of the complement sequence, since DAF normally down-regulates the C3 and C5 convertases responsible for this amplification (MEDOF et al. 1984), and (b) markedly enhanced assembly of the membrane attack complexes (MACS), since MIRL normally down-regulates this step (HOLGUIN et al. 1989b; ROLLINS and SIMS 1990). The result is that, for the effective production of a single effective transmembrane MAC, far less complement must be activated on these cells than on normal cells (Table 2) (ROSSE et al. 1974b; ROUAULT et al. 1978).

Of these two proteins, MIRL is the more important in the defense against activated complement. Complete inhibition of DAF activity on normal red cells by monoclonal or polyclonal antibodies results in cells much like PNH II cells in their sensitivity to complement (MEDOF et al. 1987), whereas similar inhibition of MIRL results in cells much like PNH III cells. Further, when MIRL is added to the membranes of PNH III cells, they become nearly normal in their sensitivity to complement (ROSSE 1990b).

PNH II cells and the other cells of intermediate sensitivity express coordinately diminished amounts of DAF and MIRL (and acetylcholinesterase) (ROSSE et al. 1991); the amount of these proteins ranges from less than 10% of normal to more than 50%. It was originally thought that PNH II cells lacked DAF but not the regulatory protein of the MAC (MIRL was not known at the time). However, it is now clear that the increased sensitivity to complement of PNH II cells appears to be primarily due to the fact that this diminished amount of DAF is insufficient to regulate the amplification steps, since the addition of purified DAF to the membrane renders the cells normal in sensitivity to complement. However, even the diminished amount of MIRL is sufficient to protect the cells from a great deal of lysis by inhibition of the formation of the MAC.

Although other mechanisms for the in vivo lysis of PNH red cells have been proposed, it now seems almost certain that the lysis is due to activated complement. In general, the amount of lysis is related to the proportion of abnormal cells and to their degree of abnormality. Thus, patients with more than 60% PNH III cells will have hemoglobinuria frequently; those with less than 20%

Table 2. Relative efficiency of complement components in producing lysis of PNH and normal human erythrocytes

	Normal	PNH
Fraction anti-I adsorbed	0.75	0.82
Molecules C1/molecule anti-I	0.10	0.12
Molecules C4/molecule C1	12	12.5
Molecules C3/molecule C4	5.32	31.2
Lytic events per C3 ($\times 10^{-6}$)[a]	5.4	24.2
Membrane attack complexes at at lysis[a]	22.0×10^3	2.0×10^3

[a] When $Z = 1.0$ (where $Z = -\ln(1-y)$ and $y =$ the fraction of cells lysed, a mean of one fatal event has occurred on each cell. Data from ROSSE et al. 1974b and ROUAULT et al. 1978

will have it almost not at all. Patients with a very high proportion of PNH II cells will almost never have hemoglobinuria. This difference is reflected in the fact that the mean life span for PNH III cells is about 8 days, whereas that of PNH II erythrocytes is about 45 days (Rosse 1971).

When paroxysms occur, they are often in response to the activation of complement. As first noted by Gull (1866), both viral and bacterial infections can cause hemoglobinuria. Incompatible transfusion, with the subsequent activation of complement, may result in about of hemoglobinuria (Dacie and Firth 1943). Even the nocturnal aspect of the hemoglobinuria has been ascribed to the activation of complement by bacterial endotoxin, the serum level of which varies in a circadian rhythm parallel to the hemoglobinuria, reaching a peak during sleeping hours.

So far as can be determined, the hemolysis is intravascular, as would be expected from the destruction of cells by complement. This results in the depletion of haptoglobin with subsequent glomerular filtration of the excess hemoglobin. A proportion of the filtered hemoglobin is reabsorbed by the proximal tubules where a large load of hemoglobin in the filtrate may interfere with the reabsorption of other substances, such as glucose and amino acids, leading to the syndrome associated with Fanconi's renal disease (Clark et al. 1981).

In the cells of the proximal tubule, hemoglobin is catabolized and the iron that is liberated is retained within the cells. This causes the kidneys to be laden with hemosiderin, sometimes in sufficient quantity to set off the electromagnetic devices in airports. This excess iron may, on occasion, interfere with renal function, sometimes causing proximal renal acidosis and sometimes perhaps contributing to renal failure, which is relatively common late in the disease (Clark et al. 1981).

As the cells of the kidney are shed, the iron they contain is detected as hemosiderin. Almost all patients with PNH have "perpetual hemosiderinuria" (Marchiafava 1928) and may lose as much as 20 mg per day, ten times the normal total body excretion, without any sign of hemoglobinuria (Sears et al. 1966). This excessive loss of iron frequently leads to iron deficiency.

4 Thrombosis

Patients with PNH are markedly prone to thrombosis of veins in unusual sites, including the hepatic veins (Peytremann et al. 1972; Hartmann et al. 1980; Horler et al. 1970), the cerebral veins (Johnson et al. 1970), the abdominal and splanchnic veins (Blum and Gardner 1966; Grossman and McDermott 1974), and the veins of the skin (Rietschel et al. 1978). So far as is known, arterial occlusion is rare. Venous thrombosis may affect the function of the organs involved and is not infrequently the cause of death in PNH.

The reason for the increased tendency to thrombosis is not known. The platelets, like the red cells, are abnormal in that they lack the GPI-linked membrane proteins (NICHOLSON-WELLER et al. 1983b; DEVINE et al. 1987a). It has been suggested that the activation of complement results in platelet aggregation (DIXON and ROSSE 1977), but when the studies were done more carefully, using purified components of complement rather than crude fractions of serum, this proved not to be the case (BRYANT et al. 1988).

The platelets in PNH appear to be extraordinarily sensitive to aggregation by thrombin; 1/10–1/100 the amount of thrombin needed to cause aggregation of normal platelets is sufficient to aggregate these platelets. They are not more readily aggregated by any of the other agonists used in such studies (ADP, thromboxane A_2, arachidonic acid, etc.). The reason for this unusual susceptibility to thrombin is not clear but does not appear to be related to the deficiency of DAF or MIRL.

Another mechanism by which the defect in PNH platelets might lead to excessive thrombosis has been proposed. Nascent membrane attack complexes (C5b-9) on normal platelets cause the membrane to vesiculate, which concentrates the complexes and removes them by pinching off of the vesicles; the vesicles that are formed in this way exhibit excess acidic phospholipids and thus provide a site for the formation of the prothrombinase complex of the coagulation system (factors V, X, and VIII) (SIMS et al. 1988). When MIRL is inhibited by antibody, the formation of these vesicles is markedly enhanced (SIMS et al. 1989). Since PNH platelets lack MIRL, it is possible that vesicle formation is enhanced in them as well when complement is even minimally activated. The generation of the procoagulant vesicles may be in part the cause of the propensity to thrombosis.

5 Increased Tendency of Infection

Some patients with PNH tend to have recurring bacterial infections. In some patients, this may be due to the granulocytopenia that is not infrequently seen in this disorder (DACIE 1963). In others, however, infections, particularly with unusual organisms, may occur when the white count is perfectly normal; of these, septicemia appears to be the most severe and the least well handled.

In part, this tendency may be due to an identified defect in the granulocytes, which lack the Fc (γ) III receptor (SELVARAJ et al. 1988), a molecule that may or may not be GPI-linked. This molecule binds to the Fc portion of IgG antibody molecules bound to antigens on the surface of the invading organism; the interaction results in immune adherence and phagocytosis. Although there are several other Fc receptors on other phagocytic cells, the GPI-linked form of this receptor is the major receptor on granulocytes, which are responsible for patrolling the circulating blood.

A number of GPI-anchored proteins with immunologic functions have been identified in mice, e.g., Thy-1 (TSE et al. 1985) and Qa-2 (WANECK et al. 1988), but their homologues have not been found in humans. One such human protein, lymphocyte function antigen 3 (LFA-3), has been found to exist in both the GPI-linked and the transmembrane-linked form (SELVARAJ et al. 1987); the blood cells in PNH lack the former but not the latter. This protein is important in signaling between lymphocytes and other cells because it is the receptor of the CD2 molecule (BIERER et al. 1988). Reaction of the LFA-3 molecule on monocytes results in the production of interleukin-1 (IL-1), a cytokine of the bone marrow, and its derivatives. To date, no specific defect in cellular immunity has been described in PNH.

6 Deficient Hematopoiesis

In most patients with PNH, there is evidence of deficient hematopoiesis. This may range from nearly complete aplasia of the bone marrow (DACIE and LEWIS 1961; ROSSE 1985) to defects that can only be detected using cell culture techniques which demonstrate that the bone marrow of PNH patients grows less well in vitro regardless of the apparent proliferative state in vivo (TUMEN et al. 1980; SULTAN et al. 1973). At least one-quarter of all patients with PNH have a preceding phase of hypoplastic hematopoiesis, and patients with established PNH not infrequently have episodes of severe hypoplasia. In nearly half of the patients with PNH, the deficient hematopoiesis is manifest in thrombocytopenia or granulocytopenia (DACIE 1963). The survival in the circulation of these two cell types is normal in PNH despite the lack of GPI-linked proteins (DEVINE et al. 1987a; BRUBAKER et al. 1977); thus, low cell numbers must be due to diminished production.

The cause of the defective hematopoiesis in PNH is not known. The stem cells presumably lack the GPI-linked proteins and this could affect their response to cytokines, etc. Cells such as T lymphocytes and monocytes modify and regulate hematopoiesis, and abnormalities in the surface expression of GPI-linked proteins might alter their ability to do so. This provides a fertile area of exploration.

7 Leukemia

About 1%–3% of patients with PNH acquire acute leukemia at the end of their illness (HOLDEN and LICHTMAN 1969; JENKINS and HARTMANN 1969; KAUFMANN et al. 1969; HIRSCH et al. 1981). This is nearly always myeloid in phenotype although acute lymphocytic leukemia has also been described (KATAHIRA et al. 1983).

The leukemia usually arises 4–7 years after the onset of PNH and is preceded by a preleukemic phase. When the leukemia appears, it is difficult to treat and remission is rare.

The leukemic cells are derived from the abnormal population of PNH cells (at least in one patient), since they lack the GPI-linked proteins (DEVINE et al. 1987b). However, it is not at all clear how or even if the defect in GPI anchor formation results in this complication. It is possible that more than one defect might arise in the abnormal population.

8 Conclusions

PNH, a hematologic disorder of protean manifestations, is characterized by the deficiency of GPI-anchored proteins on the affected cells of the blood. While the relationship between the symptoms of the disease can be directly related to the deficiency of certain of these proteins, this is not always true and a more thorough understanding of their function may be necessary. Further, it is likely that other GPI-linked proteins that have not been described are also deficient on the blood cells of patients with PNH, and their lack may account for some of the symptoms of the disorder. At any event, PNH provides an excellent opportunity to understand the biosynthesis and function of the GPI-linked proteins.

References

Auditore JV, Hartmann RC, Flexner JM, Balchum OJ (1960) The erythrocyte acetylcholinesterase enzyme in paroxysmal nocturnal hemoglobinuria. Arch Pathol 69: 534–543

Beck WS, Valentine WN (1951) Biochemical studies on leucocytes. II. Phosphatase activity in chronic lymphatic leukemia, acute leukemia, and miscellaneous hematologic conditions. J Lab Clin Med 38: 245–253

Bierer BE, Barbosa J, Herrmann S, Burakoff SJ (1988) Interaction of CD2 with its ligand, LFA-3, in human T cell proliferation. J Immunol 140: 3358–3363

Blum SF, Gardner FH (1966) Intestinal infarction in paroxysmal nocturnal hemoglobinuria. N Engl J Med 274: 1137–1138

Brubaker L, Essig LJ, Mengel CE (1977) Neutrophil life span in paroxysmal nocturnal hemoglobinuria. Blood 50: 657–662

Bryant PM, Hall SE, Cole JS, Greenberg CS, Rosse WF (1988) Marked sensitivity of paroxysmal nocturnal hemoglobinuria (PNH) platelets to thrombin: relationship to complement activation. Blood [Suppl] 72: 317a

Burroughs SF, Devine DV, Kaplan ME (1988) Paroxysmal nocturnal hemoglobinuria neutrophils deficient in decay accelerating factor are also deficient in alkaline phosphatase. Blood 71: 1086–1089

Carothers DJ, Hazra SV, Andreson SW, Medof ME (1990) Synthesis of aberrant decay-accelerating factor proteins by affected paroxysmal nocturnal hemoglobinuria leucocytes. J Clin Invest 85: 47–54

Chow F-L, Telen MT, Rosse WF (1986) The separation of the populations of red cells in paroxysmal nocturnal hemoglobinuria by monoclonal antibody to acetylcholinesterase. Blood 67: 893–897

Clark DA, Butler SA, Braren V, Hartmann RC, Jenkins DE Jr (1981) The kidneys in paroxysmal nocturnal hemoglobinuria. Blood 57: 83–89

Dacie JV (1963) Paroxysmal nocturnal haemoglobinuria. Proc R Soc Med 56: 587–596

Dacie JV, Firth D (1943) Blood transfusion in nocturnal haemoglobinuria. Br Med J i: 626

Dacie JV, Lewis SM (1961) Paroxysmal nocturnal hemoglobinuria: variation in clinical severity and association with bone marrow hypoplasia. Br J Haematol 7: 442–457

Dacie JV, Lewis SM (1972) Paroxysmal nocturnal hemoglobinuria, clinical manifestations, hematology and nature of the disease. Ser Hematol 5: 3–23

Dameshek W (1969) Forward and a proposal for considering paroxysmal nocturnal hemoglobinuria (PNH) as a "candidate" myeloproliferative disorder. Blood 33: 263–264

Devine DV, Siegel RS, Rosse WF (1987a) Interactions of the platelets in paroxysmal nocturnal hemoglobinuria with complement. J Clin Invest 79: 131–137

Devine DV, Gluck WL, Rosse WF, Weinberg JB (1987b) Acute myeloblastic leukemia in paroxysmal nocturnal hemoglobinuria: evidence of evolution for the abnormal PNH clone. J Clin Invest 79: 314–317

Devine DV, Rosse WF (1991) Deficiency of 5'-*ecto*nucleotidase on the lymphocytes in paroxysmal nocturnal hemoglobinuria. (to be published)

Dixon RH, Rosse WF (1977) Mechanism of complement activation on human blood platelets in vitro. J Clin Invest 59: 360–368

Grossman JA, McDermott WV Jr (1974) Paroxysmal nocturnal hemoglobinuria associated with hepatic and portal venous thrombosis. Am J Surg 127: 733–736

Gull WW (1866) A case of intermittent haematinuria, with remarks. Guy's Hosp Rep 12: 381–392

Ham TH (1937) Chronic hemolytic anemia with paroxysmal nocturnal hemoglobinuria. A study of the mechanism of hemolysis in relation to acid-base equilibrium. N Engl J Med 217: 915–917

Ham TH, Dingle JH (1939) Studies on destruction of red blood cells. II. Chronic hemolytic anemia with paroxysmal nocturnal hemoglobinuria: certain immunological aspects of the hemolytic mechanism with special reference to serum complement. J Clin Invest 18: 657–672

Hänsch GM, Schonermark S, Roelcke D (1987) Paroxysmal nocturnal hemoglobinuria type III. Lack of an erythrocyte membrane protein restricting the lysis of C5b-9. J Clin Invest 80: 7–12

Hartmann RC, Arnold AB (1977) Paroxysmal nocturnal hemoglobinuria as a clonal disorder. Annu Rev Med 28: 187–194

Hartmann RC, Luther AB, Jenkins DE Jr, Tenorio LE, Saba HI (1980) Fulminant hepatic venous thrombosis (Budd-Chiari syndrome) in paroxysmal nocturnal hemoglobinuria: definition of a medical emergency. Johns Hopkins Med J 146: 247–254

Hijmans van den Bergh AA (1911) Ictere hemolytique avec crises hemoglobinuriques. Fragilite globulaire. Rev Med 31: 63–69

Hirsch VJ, Neubach PA, Parker DM, Reese MH, Stone MJ (1981) Paroxysmal nocturnal hemoglobinuria. Termination in acute myelomonocytic leukemia and reappearance after leukemic remission. Arch Intern Med 141: 525–527

Holden D, Lichtman H (1969) Paroxysmal nocturnal hemoglobinuria with acute leukemia. Blood 33: 283–286

Holguin MH, Wilcox LA, Bernshaw NJ, Rosse WF, Parker CJ (1989a) Relationship between the membrane inhibitor of reactive lysis and the erythrocyte phenotypes of paroxysmal nocturnal hemoglobinuria. J Clin Invest 84: 1387–1394

Holguin MH, Frederick LR, Bernshaw NJ, Wilcox LA, Parker CJ (1989b) Isolation and characterization of a membrane protein from normal human erythrocytes that inhibits reactive lysis of the erythrocytes of paroxysmal nocturnal hemoglobinuria. J Clin Invest 84: 7–17

Horler AR, Shaw MT, Thompson RB (1970) Budd-Chiari syndrome as a complication of paroxysmal nocturnal hemoglobinuria. Postgrad Med J 46: 618

Jenkins DE Jr, Hartmann RC (1969) Paroxysmal nocturnal hemoglobinuria terminating in acute myeloblastic leukemia. Blood 33: 274–282

Johnson RV, Kaplan SR, Blailock ZR (1970) Cerebral venous thrombosis in paroxysmal nocturnal hemoglobinuria. Marchiafava-Micheli syndrome. Neurology 20: 681–686

Katahira J, Masako A, Oshimi K, Mizoguchi H, Okada M (1983) Paroxysmal nocturnal hemoglobinuria terminating in TdT-positive acute leukemia. Am J Hematol 14: 79–87

Kaufmann RW, Schechter GP, McFarland W (1969) Paroxysmal nocturnal hemoglobinuria terminating in acute granulocytic leukemia. Blood 22: 287–291

Kinoshita T, Medof ME, Silber R, Nussenzweig V (1985) Distribution of decay accelerating factor in the peripheral blood of normal individuals and patients with paroxysmal nocturnal hemoglobinuria. J Exp Med 162: 75–92

Kunstling TR, Rosse WF (1969) Erythrocyte acetylcholinesterase deficiency in paroxysmal nocturnal hemoglobinuria (PNH)—a comparison of the complement-sensitive and -insensitive populations. Blood 33: 607–616

Mahoney JF, Urakaze M, Hall S, DeGasperi R, Chang H-M, Warren CD, Nicholson-Weller A, Rosse WF, Yeh ETH (1991) Defective glycosylphosphatidylinositol anchor synthesis in paroxysmal nocturnal hemoglobinuria. Clin Res 39: 210a (abstract)

Marchiafava E (1928) Anemia emolitica con emosiderinuria perpetua. Policlinico, Sez Med 35: 109

May JE, Frank MM, Rosse WF (1973) Alternate complement-pathway-mediated lysis induced by magnesium. N Engl J Med 298: 705–709

Medof ME, Kinoshita T, Nussenzweig V (1984) Inhibition of complement activation on the surface of cells after incorporation of decay-accelerating factor (DAF) into their membranes. J Exp Med 160: 1558–1578

Medof ME, Gottleib A, Kinoshita T, Hall S, Silber R, Nussenzweig V, Rosse WF (1987) Relationship between decay accelerating factor deficiency, diminished acetylcholinesterase activity, and defective terminal complement pathway restriction in paroxysmal nocturnal hemoglobinuria erythrocytes. J Clin Invest 80: 165–174

Meri S, Morgan BP, Davies A, Daniels RH, Olavesen MG, Waldmann H, Lachmann PJ (1990) Human protectin (CD59), an 18 000–20 000 MW complement lysis restricting factor, inhibits C5b-8 catalyzed insertion of C9 into lipid bilayers. Immunology 71: 1–9

Nicholson-Weller A, March JP, Rosenfeld SI, Austen KF (1983a) Affected erythrocytes of patients with paroxysmal nocturnal hemoglobinuria are deficient in the complement regulatory protein, decay accelerating factor. Proc Natl Acad Sci USA 80: 5430–5434

Nicholson-Weller A, Spicer DB, Austen KF (1983b) Deficiency of the complement regulatory protein "decay accelerating factor," on membranes of granulocytes, monocytes, and platelets in paroxysmal nocturnal hemoglobinuria. N Engl J Med 312: 1091–1097

Pangburn MK, Schreiber RD, Muller-Eberhard HJ (1983) Deficiency of an erythrocyte membrane protein with complement regulatory activity in paroxysmal nocturnal hemoglobinuria. Proc Natl Acad Sci USA 80: 5430–5434

Peytremann R, Rhodes RS, Hartmann RC (1972) Thrombosis in paroxysmal nocturnal hemoglobinuria (PNH) with particular reference to progressive diffuse hepatic venous thrombosis. Ser Haematol 5: 115–136

Rietschel RL, Lewis CW, Simmons RA, Phyliky RL (1978) Skin lesions in paroxysmal nocturnal hemoglobinuria. Arch Dermatol 114: 560–563

Rollins SA, Sims PJ (1990) The complement-inhibitory activity of CD59 resides in its capacity to block incorporation of C9 into membrane C5b-9. J Immunol 144: 3478–3483

Rosse WF (1971) The line-span of complement-sensitive and -insensitive red cells in paroxysmal nocturnal hemoglobinuria. Blood 37: 556–562

Rosse WF (1980) Paroxysmal nocturnal hemoglobinuria—present status and future prospects. West J Med 132: 219–228

Rosse WF (1985) Paroxysmal nocturnal hemoglobinuria in aplastic anemia. Clin Haematol 14: 105–125

Rosse WF (1990a) Phosphatidylinositol-linked proteins and paroxysmal nocturnal hemoglobinuria. Blood 75: 1595–1601

Rosse WF (1990b) Paroxysmal nocturnal hemoglobinuria. In: Clinical immunohematology: basic concepts and clinical applications. Blackwell, Boston, pp 593–648

Rosse WF, Dacie JV (1966) Immune lysis of normal human and paroxysmal nocturnal hemoglobinuria red blood cells. I. The sensitivity of PNH red cells to lysis by complement and specific antibody. J Clin Invest 45: 736–748

Rosse WF, Parker CJ (1986) Paroxysmal nocturnal haemoglobinuria. Clinics in Haemat 14: 105–125

Rosse WF, Adams JP, Thorpe AM (1974a) The population of cells in paroxysmal nocturnal hemoglobinuria of intermediate sensitivity to complement lysis—significance and mechanism of increased immune lysis. Br J Haematol 28: 181–190

Rosse WF, Logue GL, Adams J, Crookston JH (1974b) Mechanisms of immune lysis of the red cells in hereditary erythroblastic multinuclearity with a positive acidified serum test and paroxysmal nocturnal hemoglobinuria. J Clin Invest 53: 31–43

Rosse WF, Hoffman S, Campbell M, Borowitz M, Moore JO, Parker CJ (1991) The erythrocytes in paroxysmal nocturnal hemoglobinuria of intermediate sensitivity to complement lysis. Br J Haematol (to be published)

Rouault TA, Rosse WF, Bell S, Shelburne J (1978) Differences in the terminal steps of complement lysis of normal and paroxysmal nocturnal hemoglobinuria red cells. Blood 51: 325–330

Sears DA, Anderson PR, Foy AL, Williams HL, Crosby WH (1966) Urinary iron excretion and renal metabolism of hemoglobin in hemolytic diseases. Blood 28: 708–725

Selvaraj P, Dustin ML, Silber R, Low MG, Springer TA (1987) Deficiency of lymphocyte function-associated antigens 3 (LFA-3) in paroxysmal nocturnal hemoglobinuria. Functional correlates and evidence for a phosphatidylinositol membrane anchor. J Exp Med 166: 1011–1025

Selvaraj P, Rosse WF, Silber R, Springer TA (1988) The major Fc receptor in blood has a phosphotidylinositol anchor and is deficient in paroxysmal nocturnal hemoglobinuria. Nature 333: 565–567

Simmons DL, Tan S, Tenen DG, Nicholson-Weller A, Seed B (1989) Monocyte antigen CD14 is a phospholipid anchored membrane protein. Blood 73: 284–289

Sims PJ, Faioni EM, Wiedmer T, Shattil SJ (1988) Complement proteins C5b-9 cause release of membrane vesicles from the platelet surface that are enriched in the membrane receptor for coagulation factor Va and express prothrombinase activity. J Biol Chem 263: 18205–18212

Sims PJ, Rollins SA, Wiedmer T (1989) Regulatory control of complement on blood platelets. Modulation of platelet procoagulant responses by a membrane inhibitor of the C5b-9 complex. J Biol Chem 264: 17049–17057

Sirchia G, Lewis SM (1975) Paroxysmal nocturnal haemoglobinuria. Clin Haematol 4: 199–299

Stafford HA, Tykocinski ML, Lublin DM, Holers VM, Rosse WF, Atkinson JP, Medof ME (1988) Normal polymorphic variations and transcription of the decay accelerating factor gene in paroxysmal nocturnal hemoglobinuria cells. Proc Natl Acad Sci USA 85: 880–884

Sultan C, Marquet M, Joffroy Y (1973) Etude de dysmyelopoieses acquises indipathiques en culture de moelle in vitro. Nouv Rev Fr Hematol 13: 431–436

Tse AGD, Berklay AN, Watts A, Williams AF (1985) A glycophospholipid tail at the carboxyl terminus of the Thy-1 glycoprotein of neurons and thymocytes. Science 230: 1003–1008

Tumen J, Kline LB, Fay J, Scullin D, Reisner E, Rosse WF, Huang A (1980) Complement sensitivity of paroxysmal nocturnal hemoglobinuria bone marrow cells. Blood 55: 1040–1046

Van der Schoot CE, Huizinga TWJ, Gadd S, Majdic O, Wijmans R, Knapp W, von dem Borne AEGK (1989) Identification of three novel PI-linked proteins on granulocytes. In: Knapp W, Dorken B, Gilks WR, Rieber EP, Schmidt RE, Stein H, von dem Borne AEGK (eds) Leucocyte typing IV: white cell differentiation antigens. Oxford University Press, Oxford, pp 887–891

Waneck Cl, Sherman PW, Kincade PW, Low MG, Flavell RA (1988) Molecular mapping of signals in the Qa-2 antigen required for attachment of the phosphatidylinositol membrane anchor. Proc Natl Acad Sci USA 85: 577–581

Perforin: Structure, Function, and Regulation

E. R. PODACK

1 Introduction 175
2 Physicochemical and Functional Properties of Perforin 176
3 Sequences of Murine and Human Perforins 178
4 Homology of Perforin to Complement Proteins: The Perforin Family 178
5 Lack of Protection of Host Cells from Homologous Perforin 179
References 180

1 Introduction

The mechanism of lymphocyte-mediated cytolysis is controversial (see recent reviews BLEACKLEY et al. 1988; BRUNET et al. 1988; CLARK et al. 1988; MUELLER et al. 1988; MÜLLER-EBERHARD 1988; MUNGER et al. 1988; NAGLER-ANDERSON et al. 1988; SITKOVSKY 1988; YOUNG et al. 1988; PODACK et al. 1988; JENNE and TSCHOPP 1988a, b; BLEACKLEY 1988; HERSHBERGER et al. 1988; STEVENS et al. 1988; TSCHOPP and NABHOLZ 1990; PODACK et al. 1991). The main contentions concern the question whether secretory functions of the killer cell, as proposed by the vectorial granule secretion model of cytotoxicity (PODACK 1985; HENKART 1985), are required for target cell lysis and whether secretory models of cytotoxicity account for most of the cytotoxic activity. Other open questions are related to the role of DNA degradation (apoptosis, programmed cell death) in target cell lysis (RUSSEL et al. 1982; RUSSEL 1983; DUKE et al. 1983; LOVELL and MARTZ 1987) and how this process is mediated molecularly. No molecular tools are available at the present time to assess the importance, or even existence, of the putative secretion-independent pathway of lymphocyte-mediated cytolysis. Similarly, since the mechanism of DNA degradation is not understood, it is difficult to quantitate its role in cytolysis.

The cytolytic activity of perforin, isolated from the granules of cytolytic T cells and natural killer (NK) cells, is undisputed (PODACK and DENNERT 1983; DENNERT and PODACK 1983; MILLARD et al. 1984; PODACK and KONIGSBERG 1984; YOUNG et al.

Department of Microbiology and Immunology, University of Miami, School of Medicine, P.O. Box 016090 (R138), Miami, FL 33103, USA

1986a; MASSON and TSCHOPP 1985; MASSON et al. 1985; PODACK et al. 1985; ZALMAN et al. 1986; PODACK 1987; YOUNG et al. 1986b). The mechanism of perforin-mediated lysis requires the vectorial secretion of perforin, either alone or in association with the cytoplasmic granules of the killer cell, onto the target membrane following recognition and conjugate formation.

Perforin has recently been isolated and characterized, including sequence determination by cDNA cloning (LICHTENHELD et al. 1988; SHINKAI et al. 1988; LOWREY et al. 1989; KWON et al. 1989). The tools generated by this work are being used for a critical assessment of the role of perforin, as opposed to that of other molecules, in lymphocyte-mediated cytolysis. In this review, the current state of knowledge of the structure and function of perforin and its relationship to the channel formers of complement and to complement regulatory proteins will be summarized. The properties of other granule proteins, granzymes and proteoglycans, have recently been reviewed (JENNE and TSCHOPP 1988a, b; STEVENS et al. 1988; BLEACKLEY 1988; HERSHBERGER et al. 1988) and will not be described here.

2 Physicochemical and Functional Properties of Perforin

Perforin is a glycoprotein with an apparent molecular weight of 70 000 upon reduction and SDS-polyacrylamide gel electrophoresis (MASSON and TSCHOPP 1985; PODACK et al. 1985; HENKART 1985; YOUNG et al. 1986). It binds to heparin with high affinity and to anion exchange resins at neutral pH. Perforin is localized in the cytoplasmic granules of cytolytic T cells and NK cells (GROSCURTH et al. 1987; PETERS et al. 1989), where it is in all likelihood associated with the granule proteoglycan chondroitin sulfate A (MCDERMOTT et al. 1985). Owing to their content of perforin, isolated cytoplasmic granule are highly cytolytically active in the presence of Ca^{2+} (MILLARD et al. 1984; PODACK and KONIGSBERG 1984; YOUNG et al. 1986; ALLBRITTON et al. 1988; CRIADO et al. 1985; MASSON et al. 1985). In granule preparations lacking cytolytic activity, despite their content of perforin, lytic activity can be generated by salt extraction (LICHTENHELD et al. 1988), indicating an ionic interaction of perforin with granule proteoglycans.

Isolated perforin is a highly cytolytic molecule. Cytolysis is mediated by the Ca^{2+}-dependent polymerization of perforin, within the attacked membrane, into a transmembrane channel of varying diameter. According to electron microscopical and electrophysiological evidence, the maximal channel size is approximately 16 nm (PODACK and DENNERT 1983; YOUNG et al. 1986; BLUMENTHAL et al. 1984; CRIADO et al. 1985). This channel is formed by a tubular (cylindrical) complex of approximately 20 perforin protomers, as estimated from electron micrographs. The polyperforin tubule is 16 nm long and inserted 4 nm deep into the lipid bilayer of the membrane. Smaller transmembrane channels are formed when fewer than 20 perforin molecules form an incompletely closed polyperforin

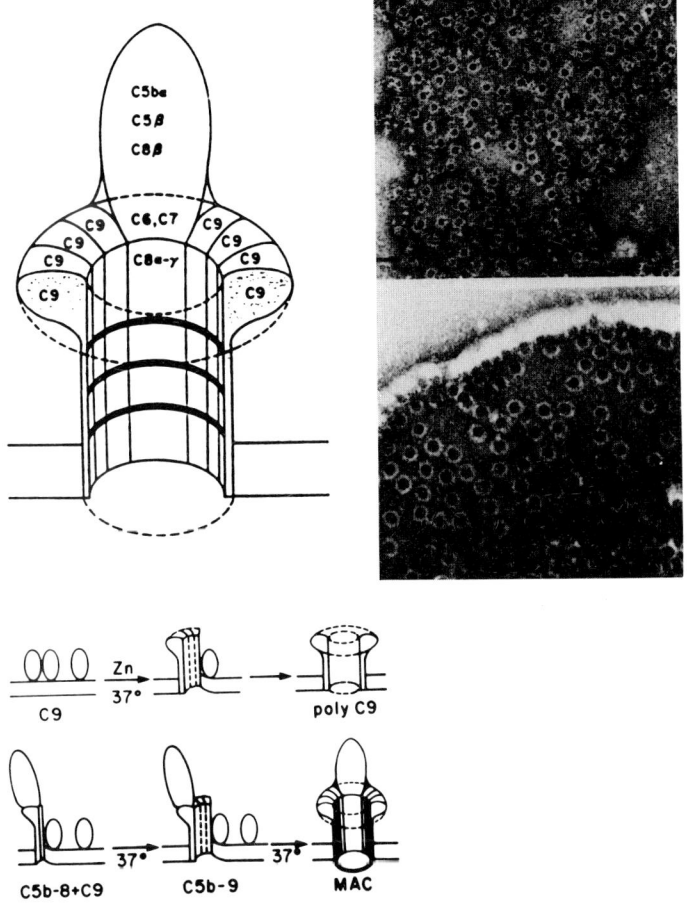

Fig. 1. Electron micrographs comparing membrane lesions formed by complement (*upper panel*) and by perforin (*lower panel*) and schematic representation of membrane attack complex (MAC), polyC9 and polyperforin assembly

complex. Although not directly demonstrated so far, the theoretical limit is a perforin dimer with a putative functional channel diameter of 1–2 nm. Transmembrane channel formation by oligomeric complexes is explained by the putative structure of the membrane spanning domain, which is postulated to be hydrophobic on one face and hydrophilic on the other. Such a structure could be formed by an amphipathic helix or an amphipathic β-pleated sheet.

Perforin polymerization is mediated by Ca^{2+} or Sr^{2+} ions but not by Mg^{2+} or Mn^{2+}. In the presence of Ca^{2+}, perforin, at low temperature (4°C), binds reversibly to natural membranes without causing lysis (YOUNG et al. 1987; YUE et al. 1987; TSCHOPP et al. 1989). Addition of EDTA allows the dissociation of perforin in active form from the membrane. At elevated temperature, membrane-associated perforin irreversibly inserts and polymerizes to a lytic transmembrane channel that cannot be dissociated by EDTA. These functional properties of perforin are quite analogous to those of C9 (Fig. 1), which reversibly binds to C5b-8 in the cold and, at elevated temperature, undergoes restricted unfolding and polymerization with the simultaneous appearance of hydrophobicity (for review see PODACK 1986) (PODACK and TSCHOPP 1982; TSCHOPP et al. 1982, 1985; PODACK et al. 1982). A unique property of perforin is its high affinity for phosphorylcholine (TSCHOPP et al. 1989; YUE et al. 1987), which acts as a Ca^{2+}-dependent receptor and thereby initiates its membrane insertion and polymerization. In the case of C9, C5b-8 is required; the interaction is Ca^{2+}-independent and proceeds in the presence of EDTA. However, C9 binds one molecule of Ca^{2+} with high affinity in the NH_2-terminal part of the molecule (THIELENS et al. 1988). Bound Ca^{2+} appears to stabilize C9. It will be interesting to determine whether perforin also is capable of binding Ca^{2+} and whether within cytolytic granules it is complexed with Ca^{2+}.

3 Sequences of Murine and Human Perforins

Perforin cDNA from murine and human cytolytic lymphocytes was recently cloned and sequenced (SHINKAI et al. 1988; LOWREY et al. 1989; LICHTENHELD et al. 1988). On the protein level the two are 67% identical. The mature proteins consist of 534 amino acids and are preceded by a 20 (murine) or 21 (human) amino acid long leader peptide, consistent with the localization of perforin in membrane enclosed granules. Perforin has two potential N-linked glycosylation sites which, however, are not in the same position in the murine and human proteins.

The coding region of murine perforin is preceded by 148 untranslated nucleotides that do not contain typical ribosomal binding sequences. The stop codon is followed by approximately 1100 nucleotides of 3' untranslated sequence. The total length of the sequenced cDNA is 2920 base pairs. This length is close to the observed length of the mRNA transcript of perforin (2.9 kb) detected in northern blots of perforin expressing cells.

4 Homology of Perforin to Complement Proteins: The Perforin Family

Even before the sequence of perforin was established it was evident from functional (PODACK and DENNERT 1983; PODACK and TSCHOPP 1982), morphological (PODACK 1984, 1986), immunological comparisons (PODACK 1987; YOUNG et al. 1986; ZALMAN et al. 1985) that perforin and C9 may be related proteins. The sequence comparison fully confirmed these predictions. The central part of perforin, containing the putative membrane binding region and a cysteine-rich domain typical for the epidermal growth factor (EGF) receptor precursor, is homologous to similar regions in C9 (DiSCIPIO et al. 1984; STANLEY et al. 1985), C8 (RAO et al. 1987; HAEFLIGER et al. 1987; HOWARD et al. 1987), C7 (DiSCIPIO et al. 1988), and C6 (CHAKRAVARTI et al. 1989; HAEFLIGER et al. 1989). The NH_2-terminal 44 and the COOH-terminal 124 amino acids of perforin are not homologous to the complement proteins. Compared to C9, the membrane binding domain of perforin is thus shifted by 120 amino acids towards the NH_2-terminal. This situation makes the model for membrane insertion of C9, as proposed by STANLEY (1988), rather unlikely for perforin.

It is of interest to note that the complement proteins, in addition to the two central domains of perforin, have evolved three additional homology domains. The implication of this may be that the three domains function in the interaction of the C6, C7, C8 and C9 proteins in forming the heteropolymeric tubule of the membrane attack complex (MAC) of complement (PODACK 1984).

Evolutionarily it is likely that a perforin-like molecule gave rise to the complement proteins of the MAC of complement. We therefore propose to name the members of this protein family as the perforin family (PODACK 1987); this terminology also contains the functional characteristics of this family, namely the ability to perforate membranes.

5 Lack of Protection of Host Cells from Homologous Perforin

Complement is under strict control to prevent lysis of host cells since its primary targets are microbial organisms. The primary targets of cytotoxic T cells, however, are host cells when they become altered. Host cell lysis is the primary host defense against virus infection. For protection of host cells from accidental complement lysis, host membrane-associated proteins have evolved named DAF (decay accelerating factor) (for review see MEDOF 1988), HRF (homologous restriction factor) (for review see HÄNSCH 1988), and protectin (CD59) (SUGITA et al. 1988; HOLGUIN et al. 1989; OKADA et al. 1989; DAVIES et al. 1989), whose unique function it is to inactivate specific complement steps in the cascade. These proteins are all attached via a glycolipid anchor in the host cell membrane. DAF acts at the level of C3b and decays or prevents formation of the C3, C5 convertase. HRF prevents C8 and C9 polymerization by interacting on the host membrane with accidentally inserted C5b-7 complexes. Protectin interferes with the binding of C7 and C8 binding to C5b-6 and with polymerization of C9.

Based on the structural and functional similarities of perforin with C9, a role for these factors in restricting complement lysis was also suspected for perforin and cytotoxic T cell lysis. Indeed, reports by ZALMAN et al. (1987, 1988) suggested self-protection of T cell from perforin via T cell HRF. In addition, a soluble form of HRF was found in isolated cytolytic granules (reviewed in MULLER-EBERHARD 1988). More recent studies, however, failed to support these conclusions. Murine and human perforin lyse homologous and heterologous erythrocytes (LICHTENHELD et al. 1988; JIANG et al. 1988) and nucleated cells equally well. In addition, it was shown by HOLLANDER et al. (1989) and KRÄHENBÜHL et al. (1989) that the absence of glycolipid-anchored proteins on erythrocytes or on nucleated target cells does not increase their susceptibility to lysis by protein or cytolytic granules, cytotoxic T cells or large granular lymphocytes (NK cells), or antibody-dependent cell-mediated cytotoxicity (ADCC). Cells lacking glycolipid-anchored proteins are dramatically more susceptible to C5b-9-mediated lysis than normal cells.

Human protectin (CD59) is not able to protect cells from human (or murine) perforin-mediated lysis even though protectin protects these cells from human C5b-9. Heterologous cells, such as guinea pig erythrocytes, are protected from lysis by human C5b-9, but not from lysis by human perforin, by as little as 2000 human protectin molecules inserted into their membrane. Antibodies to human protectin abolish the protective effect towards C5b-9 lysis but have no influence on perforin-mediated lysis (MERI et al. 1990).

The accumulated evidence thus does not support a role for the complement restricting factors in cell-mediated cytolysis. This appears to be in logical agreement with the task of killer cells to attack host cells once the latter have been altered by pathological processes. The killer cells themselves, however, have to be protected from perforin and other lytic factors if they are to continue to lyse more than one target cell. It has indeed been shown that cytotoxic T cells are relatively resistant to lysis by other cytotoxic T cells (KRANZ and EISEN 1987) and to perforin- and granule-mediated lysis. The molecular basis of this resistance is not known. PETERS et al. (1989, 1990) suggested that perforin may be secreted encapsulated in targeted vesicles that unidirectionally home in on target cells by virtue of the T cell receptor and other adhesion molecules on the capsule. If this hypothesis is correct, the killer cell would not be endangered by its own cytolytic proteins.

Acknowledgement. This work was supported by NIH ≠ CA39201 and ACS ≠ IM-556.

References

Allbritton NL, Verret CR, Wolley RC, Eisen HN (1988) Calcium ion concentrations and DNA fragmentation in target cell destruction by murine cloned cytotoxic T lymphocytes. J Exp Med 167: 514

Bleackley RC (1988) The isolation and characterization of two cytolytic lymphocyte specific serine protease genes. In: Podack ER (ed) Cytotoxic effector mechanisms. Springer, Berlin Heidelberg New York, p 67 (Current topics in microbiology and immunology, vol 140)

Bleackley RC, Lobe CG, Duggan B, Ehrman N, Fregeau C, Meier M, Letellier M, Caliopi H, Shaw J, Paetkau V (1988) The isolation and characterization of a family of serine protease genes expressed in activated cytotoxic lymphocytes. Immunol Rev 103: 5

Blumenthal RP, Millard PJ, Henkart MP, Reynolds CW, Henkart PA (1984) Liposomes as targets for granule cytolysin from cytotoxic large granular lymphocyte tumors. Proc Natl Acad Sci USA 81: 5551

Brunet J-F, Denizot F, Goldstein P (1988) A differential molecular biology search for genes preferentially expressed in functional T-lymphocytes: the CTLA genes. Immunol Rev 103: 21

Chakravarti DN, Chakravarti B, Parra CA, Muller-Eberhard HJ (1989) Structural homology of complement protein C6 with other channel-forming proteins of complement. Proc Natl Acad Sci USA 86: 2799

Clark W, Ostergaard H, Gormann K, Torbett B (1988) Molecular mechanism of CTL-mediated lysis: a cellular perspective. Immunol Rev 103: 37

Criado M, Lindstrom JM, Anderson CG, Dennert G (1985) Cytotoxic granules from killer cells: specificity of granules and insertion of channels of defined sizes into target membranes. J Immunol 135: 4245

Davies A, Simmons D, Hale G, Harrison RA, Tighe H, Lachmann PJ, Waldmann H (1989) CD59, and Ly-6 like protein expressed in human lymphoid cells, regulated the action of the complement membrane attack complex on homologous cells. J Exp Med 170: 637

Dennert G, Podack ER (1983) Cytolysis by H2-specific killer cells. J Exp Med 157: 1483

DiScipio RG, Gehring MR, Podack ER, Kan CC, Hugli TE, Fey GH (1984) Nucleotide sequence of cDNA and derived amino acid sequence of human complement component C9. Proc Natl Acad Sci USA 81: 7289

DiScipio RG, Chakravarti DN, Muller-Eberhard HJ, Fey GH (1988) Structure of human C7 and the C5b-7 complex. J Biol Chem 263: 549

Duke RC, Chervenak R, Cohen JJ (1983) Endogenous endonuclease-induced DNA fragmentation: An early event in cell-mediated cytolysis. Proc Natl Acad Sci USA 80: 6361

Groscurth P, Qiao BY, Podack ER, Hengartner H (1987) Cellular localization of perforin 1 in murine cloned cytotoxic lymphocytes. J Immunol 138: 2749

Groscurth P, Qiao B-Y, Podack ER, Hengarnter H (1987) Cellular localization of perforin 1 in murine cloned cytotoxic T lymphocytes. J Immunol 138: 2749–2752

Haefliger JA, Tschopp J, Nardelli D, Wahli W, Kocher HP, Tosi M, Stanley KK (1987) Complementary DNA cloning of complement C8 beta and its sequence homology to C9. Biochemistry 26: 3551

Haefliger J-A, Tschopp J, Vial N, Jenne DE (1989) Complete primary structure and functional characterization of the sixth component of the human complement system: identification of the C5b-binding domain in complement C6. J Biol Chem 264: 18041–18051

Hänsch GM (1988) The homologous species restriction of the complement attack: structure and function of the C8 binding protein. In: Podack ER (ed) Cytoxic effector mechanisms. Springer, Berlin Heidelberg New York, p 109 (Currennt topics in microbiology and immunology, vol 140)

Henkart PA (1985) Mechanism of lymphocyte mediated cytotoxicity. Annu Rev Immunol 3: 31

Hershberger RJ, Mueller C, Gershenfeld HK, Weissman IL (1988) A serine protease encoding gene that marks activated cytotoxic T-cells in vivo and in vitro. In: Podack ER (ed) Cytotoxic effector mechanisms. Springer, Berlin Heidelberg New York, pp 81–90 (Current topics in microbiology and immunology, vol 140)

Holguin MH, Fredrick LR, Bernshaw NJ, Wilcox LA, Parker CJ (1989) Isolation and characterization of a membrane protein from normal human erythrocytes that inhibits reactive lysis of the erythrocytes of paroxysmal nocturnal hemoglobinuria. J Clin Invest 84: 7

Hollander N, Shin M, Rosse WR, Springer TA (1989) Distinct restriction of complement and cell-mediated lysis. J Immunol 142: 3913

Howard OMZ, Rao AG, Sodetz JM (1987) Complementary DNA and derived amino acid sequence of the beta subunit of complement protein C8: identification of a close structural and ancestral relationship to the alpha subunit and C9. Biochemistry 26: 3565

Jenne D, Tschopp J (1988a) Granzymes, a family of serine protease released from granules of cytolytic T lymphocytes upon cell receptor stimulation. Immunol Rev 103: 53

Jenne DE, Tschopp J (1988b) Granzymes: a family of serine esterases in granules of cytolytic lymphocytes. In: Podack ER (ed) Cytotoxic effector mechanisms. Springer, Berlin Heidelberg New York, p 33 (Current topics in microbiology and immunology, vol 140)

Jiang S, Pereschini PM, Zychlinsky A, Liu C-C, Perussia B, Young JD-E (1988) Resistance of cytolytic lymphocytes to perforin-mediated killing. Lack of correlation with complement-associated homologous species restriction. J Exp Med 169: 2207

Krähenbühl OP, Peter HH, Tschopp J (1989) Absence of homologous restriction factor does not affect CTL-mediated cytolysis. Eur J Immunol 19: 217

Kranz DM, Eisen HN (1987) Resistance of cytotoxic T lymphocytes to lysis by a clone of cytotoxic T lymphocytes. Proc Natl Acad Sci USA 84: 3375

Kwon BS, Wakulchik M, Liu CC, Pereschini PM, Trapani JA, Haq AK, Kim Y, Young JD-E (1989) The structure of the mouse lymphocyte pore-forming protein perforin. Biochem Biophys Res Commun 158: 1–10

Lichtenheld MG, Lu P, Olsen KJ, Lowrey DM, Hameed A, Hengartner H, Podack ER (1988) Structure and function of human perforin. Nature 335: 448

Lovell DM, Martz E (1987) The degree of CTL induced DNA solubilization is not determined by the human vs mouse origin of the target cell. J Immunol 138: 3695

Lowrey DM, Olsen KJ, Lichtenheld MG, Aebischer T, Rupp F, Hengartner H, Podack ER (1989) Cloning and analysis and expression of murine perforin 1 cDNA, a component of cytolytic T-cell granules with homology to complement component C9. Proc Natl Acad Sci USA 86: 247

Masson D, Tschopp J (1985) Isolation of a lytic, pore-forming protein (perforin) from cytolytic T-lymphocytes. J Biol Chem 260: 9069

Masson D, Corthesy P, Nabholz M, Tschopp J (1985) Appearance of cytolytic granules upon induction of cytolytic activity in CTL-hybrids. EMBO J 4: 2533

McDermott RP, Schmidt RE, Caulfield JP, Hein A, Bartley GT, Ritz J, Schlossman SF, Austen KF, Stevens RL (1985) Proteoglycans in cell mediated cytotoxicity. Identification, localization and exocytosis of a chondroitinsulfate proteoglycan from human cloned natural killer cells during target cell lysis. J Exp Med 162: 1771

Medof EM (1988) Decay accelerating factor and the defect of paroxysmal nocturnal hemoglobinuria. In: Podack ER (ed) Cytolytic lymphocytes and complement: effectors of the immune system. CRC Press, Boca Raton, pp 57–88

Meri S, Morgan BP, Wing M, Jones J, Davies A, Podack ER, Lachmann PJ (1990) Human protectin (CD59), and 18–20-kD homologous complement restriction factor, does not restrict perforin-mediated lysis. J Exp Med 172: 367

Millard PJ, Reynolds CW, Henkart MP (1984) Cytolytic activity of purified cytoplasmic granules from cytotoxic rat large granular lymphocyte tumors. J Exp Med 160: 75

Mueller C, Gershenfeld HK, Weissman IL (1988) Activation of CTL-specific genes during cell-mediated cytolysis in vivo: expression of HF gene analyzed by in situ hybridization. Immunol Rev 103: 73

Müller-Eberhard HJ (1988) The molecular basis of target cell killing by human lymphocytes and of killer self protection. Immunol Rev 103: 87

Munger WE, Berrebi GA, Henkart PA (1988) Possible involvement of CTL granule protease in Target cell DNA breakdown. Immunol Rev 103: 99

Nagler-Anderson C, Allbritton NL, Verret CR, Eisen HN (1988) A comparison of the cytolytic properties of CD8+ cytotoxic T lymphocytes and cloned cytotoxic T cell lines. Immunol Rev 103: 111

Okada NR, Harada T, Fujita T, Okada H (1989) A novel membrane glycoprotein capable of inhibiting membrane attack by homologous complement. Int Immunol 1: 205

Peters PJ, Geuze HJ, Van der Donk HA, Slot JW, Griffith JM, Stam NJ, Clevers HC, Borst J (1989) Molecules relevant for T cell-target cell interaction are present in cytolytic granules of human T lymphocytes. Eur J Immunol 19: 1469–1475

Peters PJ, Geuze HJ, Van der Donk HA, Borst J (1990) A new model for lethal hit delivery by cytotoxic T lymphocytes. Immunol Today 11: 38–32

Podack ER (1984) Molecular composition of the tubular structure of the membrane attack complex of complement. J Biol Chem 259: 8641

Podack ER (1985) The molecular mechanism of lymphocyte mediated tumor lysis. Immunol Today 6: 12

Podack ER (1986) Molecular mechanism of cytolysis by complement and by cytolytic lymphocytes. J Cell Biochem 30: 133

Podack ER (1987) Perforins: a family of pore forming proteins in immune cytolysis. In: Collier, Bonavide (eds) Membrane mediated cytotoxicity. Liss, New York, pp 339–352

Podack ER, Dennert G (1983) Cell mediated cytolysis: assembly of two types of tubules with putative cytolytic function by cloned natural killer cells. Nature 302: 442

Podack ER, Konigsberg PJ (1984) Cytolytic T-cell granules: Isolation, structural, biochemical and functional characterization. J Exp Med 160: 695

Podack ER, Tschopp J (1982) Polymerization of the ninth component of complement (C9): formation of poly (C9) with a tubular ultrastructure resembling the membrane attack complex of complement. Proc Natl Acad Sci USA 79: 574

Podack ER, Tschopp J, Muller-Eberhard HJ (1982) Molecular organization of C9 within the membrane attack complex of complement. J Exp Med 156: 268

Podack ER, Young JDE, Cohn ZA (1985) Isolation and biochemical and functional characterization of perforin 1 from cytolytic T-cell granules. Proc Natl Acad Sci USA 82: 8629

Podack ER, Lowrey DM, Lichtenheld MG, Olsen KJ, Aebischer T, Binder D, Rupp F, Hengartner H (1988) Structure, function and expression of murine and human perforin 1 (P1). Immunol Rev 103: 203

Podack ER, Hengartner H, Lichtenheld MG (1991) A central role of perforin in cytolysis? Annu Rev Immunol 9: 129–157

Rao AG, Howard SC, Whitehead AS, Colten HR, Sodetz JM (1987) Complementary DNA and derived amino acid sequence of the alpha subunit of human complement protein C8: evidence for the existence of a separate alpha subunit mRNA. Biochemistry 26: 3556

Russel JH (1983) Internal disintegration model of cytotoxic lymphocyte induced target damage. Immunol Rev 72: 97

Russel JH, Masakowski V, Rucinsky T, Phillips G (1982) Mechanisms of immune lysis. III. Characterization of the nature and kinetics of the cytotoxic T lymphocyte-induced nuclear lesion in the target. J Immunol 128: 2087

Shinkai Y, Takio K, Okumura K (1988) Homology of perforin to the ninth component of complement. Nature 334: 525

Sitkovsky MV (1988) Mechanistic, functional, and immunopharmacological implications of biochemical studies of antigen receptor triggered cytolytic T-lymphocyte activation. Immunol Rev 103: 127

Stanley KK (1988) The molecular mechanism of complement component C9 insertion and polymerization in biological membranes. In: Podack ER (ed) Cytotoxic effector mechanisms. Springer, Berlin Heidelberg New York, p 49 (Current topics in microbiology and immunology, vol 140)

Stanley KK, Kocher HP, Luzio JP, Jackson P, Tschopp J (1985) The sequence and topology of human complement component C9. EMBO J 4: 375–382

Stevens RL, Kamada MN, Serafin WE (1988) Structure and function of the family of proteoglycans that reside in the secretory granules of natural killer cells and other effector cell of the immune response. In: Podack ER (ed) Cytotoxic effector mechanisms. Springer, Berlin Heidelberg New York, p 93 (Current topics in microbiology and immunology, vol 140)

Sugita Y, Nakano Y, Tomita M (1988) Isolation from human erythrocytes of a new membrane protein which inhibits the formation of complement transmembrane channels. J Biochem (Tokyo) 104: 633

Thielens NM, Lohner K, Esser A (1988) Human complement protein C9 is a Calcium binding protein. J Biol Chem 263: 6665

Tschopp J, Nabholz M (1990) Perforin-mediated target cell lysis by cytolytic T lymphocytes. Annu Rev Immunol 8: 279–302

Tschopp J, Muller-Eberhard HJ, Podack ER (1982) Formation of transmembrane tubules by spontaneous polymerization of the hydrophilic complement protein C9. Nature 298: 534

Tschopp J, Podack ER, Muller-Eberhard HJ (1985) The membrane attack complex of complement: C5b-8 complex as accelerator of C9 polymerization. Proc Natl Acad Sci USA 134: 495

Tschopp J, Schafer S, Masson D, Peitsch MC, Heusser C (1989) Phosphorylcholine acts as Ca^{2+}-dependent receptor molecule for lymphocyte perforin. Nature 337: 272

Young JDE, Cohn ZA, Podack ER (1986a) The ninth component of complement and the pore-forming protein (perforin 1) from cytotoxic T-cells: Structural and functional homologies. Science 233: 184–190

Young JDE, Hengartner H, Podack ER, Cohn ZA (1986b) Purification and characterization of a cytolytic pore-forming protein from granules of cloned lymphocytes with natural killer cell activity. Cell 44: 849

Young JDE, Damiano A, DiNome MA, Leong LG, Cohn ZA (1987) Dissociation of membrane binding and lytic activities of the lymphocyte pore-forming protein (perforin). J Exp Med 165: 1371

Young JD-E, Liu C-C, Persechini PM, Cohn ZA (1988) Perforin-dependent and -independent pathways of cytotoxicity mediated by lymphocytes. Immunol Rev 103: 161

Yue CC, Reynolds CW, Henkart PA (1987) Inhibition of cytolysin activity in large granular lymphocyte granules by lipids: evidence for a membrane insertion mechanism of lysis. Mol Immunol 24: 647

Zalman LS, Brothers MA, Müller-Eberhard HJ (1985) A C9 related channel forming protein in the cytoplasmic granules of human larger granular lymphocytes. Biosci Rep 5: 1093

Zalman LS, Brothers MA, Chiu FJ, Müller-Eberhard HJ (1986) Mechanism of cytotoxicity of human large granular lymphocytes: relationship of cytotoxic lymphocyte protein to C8 and C9 of human complement. Proc Natl Acad Sci USA 83: 562

Zalman LS, Wood LM, Müller-Eberhard HJ (1987) Inhibition of antibody dependent lymphocyte cytotoxicity by homologous restriction factor incorporated into target cell membranes. J Exp Med 166: 947

Zalman LS, Brothers MA, Müller-Eberhard HJ (1988) Self protection of cytotoxic lymphocytes: a soluble form of homologous restriction factor in cytoplasmic granules. Proc Natl Acad Sci USA 83: 5262–5266

Subject Index

acidified serum test 80
AIDS 24
aminoethylisothiouronium bormide (AET) 73
animal models, MAC in 134
animals, transgenic 49
antibody-dependent cellular cytotoxicity 94, 96
– control by HRF 95
autoimmunity 48

C2, turnover of native 10
C3
– activation 47
– affinity chromatography 55
– convertases 47
C3b, tickover 46
C4 binding protein 47
C8 binding protein 87, 109
C9
– calcium binding by 126
– phosphorylation of 127
calcium
– MAC removal 121–124
– non-lethal effects 131
– oscillations 123
– release of eicosanoids 129
– release of reactive oxygen 129
– release from stores 124
cAMP
– cell lysis 118
– recovery 118
carbohydrate, N-linked complex 50
CD59 46, 61, 68, 87, 109
– relationship to HRF 87
– role in recovery 126
cDNA isoforms 51
cobra venom factor (CoF) 63, 64
cofactor activities 46, 47
cold agglutinin syndrome 48

collagen, MAC stimulation of synthesis 131
collagenase, MAC stimulation of synthesis 131
complement
– activation 134, 135
– – and platelets 101–111
– – regulator of 35, 45
– alternative pathway (APC) 46
– – activation 1, 2
– – protecting of self against 3
– classical pathway (CPC) 46, 49
– – activation of 1
– – control 3
– complexes, terminal
– – in synovial fluid 134
– – in urine 121
– lysis, homologous restriction 62, 79, 80
– receptor type 1 (CR1) 31–41, 45, 48
– – allotypes 34
– – factor I cofactor activity 37
– – functions, regulatory 36
– – isolation 34
– – molecular biology 34
– – number of CR1 per erythrocyte 39
– – sCR1 40
– – recombinant 36
– – specificity 36
– receptor type 2 (CR2) 45
– regulation, defect 165
– soluble form 49
– system 46, 49
Cromer blood group 69
cytokinase, release by MAC 130, 131
cytolysis 175
cytotoxic lymphocytes
– HRF on cell surface 95
– HRF in granules 95
cytotoxicity, cellular 94–96

decay accelerating factor (DAF)   8–29, 46, 89, 90, 93, 95, 105, 141, 142, 144–148, 150, 153, 179
– β-sheets, antiparallel   15
– binding   13
– cofactor activity   11
– cross-linking   21
– cysteines   15
– DAF-A   18
– DAF-B   18
– deficiency
– – inherited   69
– – in PNH   68, 69
– expression   19, 20
– gene   13, 14
– glycan phosphatidylinositol anchor   15–17
– glycosylation   17
– guinea pig   25
– Inab phenotype   22, 23
– intracellular pool   20
– mRNA
– – spliced   14
– – unspliced   14
– N-glycosylation   18
– northern blot analyses   15
– O-linked side chains   18
– paroxysmal nocturnal hemoglobinuria   23, 24
– rat   25
– shedding   20
– urine form   20
DNA degradation   175

ectocytosis, removal of MACs   120, 125
eicosanoids
– release by C5b-8   129
– release by MAC   128, 129
endocytosis
– clathrin-coated pits   125
– removal of MACs   120, 125
energy stores, depletion of   126
epithelial cells, sorting in polarized   155
erythrocytes   39
– mouse   25

factor H   47, 48
factor I   36, 37, 47
factor XIIIB   53

G protein system   124
glycan-phosphatidylinositol linked proteins   15–17, 164

glycoprotein Ia/IIa   103
glycosyl phosphatidylinositol (GPI)
– anchor   66, 71, 74–76, 141–157
– – anchor attachment   145–148
– – cell activation   153
– – functions   149
– – GPI anchored membrane proteins   154
– – – biosynthesis   144
– – structure   142, 143
– anchored proteins   5
gp45-70   45
guinea pigs   104

hematopoiesis, deficient   169
hemolysis   165
herpes simplex virus   49
HLA molecules   55
homologous restriction factor (HRF)   87–97, 141, 179
– ability to protect against cell-mediated cytotoxicity   96
– affinity for C8   92
– effect of antibody to HRF   89
– from cytotoxic lymphocytes   95
– from erythrocytes   89
– from urine   91
– lack of relationship to S protein   93
– relationship to C8 and C9   89
homologous restriction factor (HRF)/C8 binding protein (C8bp)
– comparison with MIRL   70
– deficiency in PNH   64, 65
– function   64, 65
– isolation   64
homologous restriction of complement lysis   62, 79, 80

immune
– adherence   33
– complexes, transport   38
immunity
– autoimmunity   48
– transplantation immunity   48
– tumor immunity   48
Inab phenotype   22, 23, 69
inhibitory proteins, role of recovery   126
intracellular signals
– MAC elimination   121–125
– non-lethal effects   131

lateral diffusion coefficients, membrane proteins   151
leukemia   169

Subject Index

long homologous repeat (LHR) 35
lupus erythematosus, systemic (SLE) 38, 39
Ly-6 72, 81, 82
lysis
– reactive 63, 64, 88
– tumor cell 24

MAC (membrane attack complex) 3, 62–65, 70, 77, 78, 80, 108, 115–135
– in animal models 134
– eicosanoids 128, 129
– history 115
– lability on cells 118
– matabolic inhibitors and lysis 117
– pathogenesis 133, 135
– reactive oxygen metabolites 128
– removal during recovery 118–121
– resistance of cells to lysis 116, 117
– rheumatoid arthritis 134, 135
– stimulation of protein synthesis 130
MCP (membrane cofactor protein) 45–57, 106, 146
– CD46 45
– gene 53
– gorilla 56
– orangutan 56
– quantification 56
– rabbit platelets 56
– on sperm 54
– on syncytial trophoblasts 56
MIRL
– expression
– – cell lines 76, 77
– – human tissues 76
– – inherited deficiency 69
– – PNH 68, 69
– functional activity
– – comparison with HRF/C8bp 70
– – complement activation, early stages 80, 81
– – endothelial cells 79
– – homologous restriction 79, 80
– – lysis, cell mediated 81
– – MAC formation, inhibition 77, 78
– – platelets 78, 79
– – T cell activation 81, 82
– isolation
– – affinity chromatography 67, 68
– – chromatographic procedure 65–68
– – from urine 74
– structural analysis
– – alternate forms 73, 74

– – glysosylation 72
– – homology with Ly-6 72
– – primary sequence 70, 71
– – RNA transcripts 73
– – sensitivity to chemical agents 72, 73
mouse erythrocytes 25
multi-hit kinetics 118
myasthenia gravis, experimental 134
myelin, effects of MAC 131

N-linked complex carbohydrate 50
N-linked sugars 50
NC (nucleated cells)
– lability of MAC 118
– resistance to lysis 116
nephritic factor 11
nephritis, experimental 134
NK cell 19
non-lethal
– attack by MACs 120
– effects of MAC 127, 128
nucleated cells see NC

O-linked sugars 50
oligodendrocyte
– calcium flux 123
– myelin synthesis 131
– vesiculation 120

parasites 25
– Schistosoma mansoni 25
– Trypanosoma cruzi 25
paroxysmal mocturnal hemoglobinuria (PNH) 4, 23, 24, 89, 106, 148, 163–170
– clinical symptoms 164
– complement regulation, defect 165
– DAF deficiency 90
– erythrocyte phenotypes 63, 68, 69
– hematopoiesis, deficient 169
– HRF deficiency 90
– HRF/C8bp deficiency 64, 65
– MIRL deficiency 66, 69, 77
– reactive lysis of 63, 64
– similarity to AET treated cells 73
– tendency to infection 168
pathogenesis and MAC 133, 135
perforin 81, 175–180
phosphatidylinositol-specific phospholipase C (PIPLC) 74, 75
platelets
– activation 102
– C1 inhibitor 104

platelets
- C1q receptor 103
- C3a 104
- C3b 107
- coagulation process 102
- and complement activation 101–111
- DAF 105
- factor D 104
- factor H 107
- membrane attack complex 108
- membrane cofactor protein (MCP) 106
- rat 107
protectin 179
protein kinase, activation by MAC 123

rat 107
RCA region 14
reactive lysis 63, 64, 88
reactive oxygen metabolites, release by MAC 128
recombinant sCR1 36
regulator of complement activation (RCA) 35, 45
rheumatoid joint, complement activation 134, 135

self/non-self discrimination 49
short consensus repeat (SCR) 15, 35, 51
species restriction 8
sugars
- N-linked 50
- O-linked 50

T cell rosetting 82
thrombosis 167
transcytosis
- clathrin-coatid pits 125
- removal of MACs 121
transgenic animals 49
transplantation immunity 48
trophoblast tissue 55
tumor cell lysis 24
tumor immunity 48

urine, MACs in 121

vaccinia virus 49
vectorial granule secretion model 175
vesicles
- removal of MACs 120
- sorting of membrane components 125
vesiculation 109
vitronectin receptor 109

xenografts 49

Current Topics in Microbiology and Immunology

Volumes published since 1987 (and still available)

Vol. 134: **Oldstone, Michael B. (Ed.):** Arenaviruses. Biology and Imminotherapy. 1987. 33 figs. VII, 242 pp. ISBN 3-540-14322-6

Vol. 135: **Paige, Christopher J.; Gisler, Roland H. (Ed.):** Differentiation of B Lymphocytes. 1987. 25 figs. IX, 150 pp. ISBN 3-540-17470-2

Vol. 136: **Hobom, Gerd; Rott, Rudolf (Ed.):** The Molecular Biology of Bacterial Virus Systems. 1988. 20 figs. VII, 90 pp. ISBN 3-540-18513-5

Vol. 137: **Mock, Beverly; Potter, Michael (Ed.):** Genetics of Immunological Diseases. 1988. 88 figs. XI, 335 pp. ISBN 3-540-19253-0

Vol. 138: **Goebel, Werner (Ed.):** Intracellular Bacteria. 1988. 18 figs. IX, 179 pp. ISBN 3-540-50001-4

Vol. 139: **Clarke, Adrienne E.; Wilson, Ian A. (Ed.):** Carbohydrate-Protein Interaction. 1988. 35 figs. IX, 152 pp. ISBN 3-540-19378-2

Vol. 140: **Podack, Eckhard R. (Ed.):** Cytotoxic Effector Mechanisms. 1989. 24 figs. VIII, 126 pp. ISBN 3-540-50057-X

Vol. 141: **Potter, Michael; Melchers, Fritz (Ed.):** Mechanisms in B-Cell Neoplasia 1988. Workshop at the National Cancer Institute, National Institutes of Health, Bethesda, MD, USA, March 23–25, 1988. 1988. 122 figs. XIV, 340 pp. ISBN 3-540-50212-2

Vol. 142: **Schüpach, Jörg:** Human Retrovirology. Facts and Concepts. 1989. 24 figs. 115 pp. ISBN 3-540-50455-9

Vol. 143: **Haase, Ashley T.; Oldstone Michael B. A. (Ed.):** In Situ Hybridization 1989. 22 figs. XII, 90 pp. ISBN 3-540-50761-2

Vol. 144: **Knippers, Rolf; Levine, A. J. (Ed.):** Transforming. Proteins of DNA Tumor Viruses. 1989. 85 figs. XIV, 300 pp. ISBN 3-540-50909-7

Vol. 145: **Oldstone, Michael B. A. (Ed.):** Molecular Mimicry. Cross-Reactivity between Microbes and Host Proteins as a Cause of Autoimmunity. 1989. 28 figs. VII, 141 pp. ISBN 3-540-50929-1

Vol. 146: **Mestecky, Jiri; McGhee, Jerry (Ed.):** New Strategies for Oral Immunization. International Symposium at the University of Alabama at Birmingham and Molecular Engineering Associates, Inc. Birmingham, AL, USA, March 21–22, 1988. 1989. 22 figs. IX, 237 pp. ISBN 3-540-50841-4

Vol. 147: **Vogt, Peter K. (Ed.):** Oncogenes. Selected Reviews. 1989. 8 figs. VII, 172 pp. ISBN 3-540-51050-8

Vol. 148: **Vogt, Peter K. (Ed.):** Oncogenes and Retroviruses. Selected Reviews. 1989. XII, 134 pp. ISBN 3-540-51051-6

Vol. 149: **Shen-Ong, Grace L. C.; Potter, Michael; Copeland, Neal G. (Ed.):** Mechanisms in Myeloid Tumorigenesis. Workshop at the National Cancer Institute, National Institutes of Health, Bethesda, MD, USA, March 22, 1988. 1989. 42 figs. X, 172 pp. ISBN 3-540-50968-2

Vol. 150: **Jann, Klaus; Jann, Barbara (Ed.):** Bacterial Capsules. 1989. 33 figs. XII, 176 pp. ISBN 3-540-51049-4

Vol. 151: **Jann, Klaus; Jann, Barbara (Ed.):** Bacterial Adhesins. 1990. 23 figs. XII, 192 pp. ISBN 3-540-51052-4

Vol. 152: **Bosma, Melvin J.; Phillips, Robert A.; Schuler, Walter (Ed.):** The Scid Mouse. Characterization and Potential Uses. EMBO Workshop held at the Basel Institute for Immunology, Basel, Switzerland, February 20–22, 1989. 1989. 72 figs. XII, 263 pp. ISBN 3-540-51512-7

Vol. 153: **Lambris, John D. (Ed.):** The Third Component of Complement. Chemistry and Biology. 1989. 38 figs. X, 251 pp. ISBN 3-540-51513-5

Vol. 154: **McDougall, James K. (Ed.):** Cytomegaloviruses. 1990. 58 figs. IX, 286 pp. ISBN 3-540-51514-3

Vol. 155: **Kaufmann, Stefan H. E. (Ed.):** T-Cell Paradigms in Parasitic and Bacterial Infections. 1990. 24 figs. IX, 162 pp. ISBN 3-540-51515-1

Vol. 156: **Dyrberg, Thomas (Ed.):** The Role of Viruses and the Immune System in Diabetes Mellitus. 1990. 15 figs. XI, 142 pp. ISBN 3-540-51918-1

Vol. 157: **Swanstrom, Ronald; Vogt, Peter K. (Ed.):** Retroviruses. Strategies of Replication. 1990. 40 figs. XII, 260 pp. ISBN 3-540-51895-9

Vol. 158: **Muzyczka, Nicholas (Ed.):** Viral Expression Vectors. 1992. 20 figs. IX, 176 pp. ISBN 3-540-52431-2

Vol. 159: **Gray, David; Sprent, Jonathan (Ed.):** Immunological Memory. 1990. 38 figs. XII, 156 pp. ISBN 3-540-51921-1

Vol. 160: **Oldstone, Michael B. A.; Koprowski, Hilary (Ed.):** Retrovirus Infections of the Nervous System. 1990. 16 figs. XII, 176 pp. ISBN 3-540-51939-4

Vol. 161: **Racaniello, Vincent R. (Ed.):** Picornaviruses. 1990. 12 figs. X, 194 pp. ISBN 3-540-52429-0

Vol. 162: **Roy, Polly; Gorman, Barry M. (Ed.):** Bluetongue Viruses. 1990. 37 figs. X, 200 pp. ISBN 3-540-51922-X

Vol. 163: **Turner, Peter C.; Moyer, Richard W. (Ed.):** Poxviruses. 1990. 23 figs. X, 210 pp. ISBN 3-540-52430-4

Vol. 164: **Bækkeskov, Steinnun; Hansen, Bruno (Ed.):** Human Diabetes. 1990. 9 figs. X, 198 pp. ISBN 3-540-52652-8

Vol. 165: **Bothwell, Mark (Ed.):** Neuronal Growth Factors. 1991. 14 figs. IX, 173 pp. ISBN 3-540-52654-4

Vol. 166: **Potter, Michael; Melchers, Fritz (Ed.):** Mechanisms in B-Cell Neoplasia 1990. 143 figs. XIX, 380 pp. ISBN 3-540-52886-5

Vol. 167: **Kaufmann, Stefan H. E. (Ed.):** Heat Shock Proteins and Immune Response. 1991. 18 figs. IX, 214 pp. ISBN 3-540-52857-1

Vol. 168: **Mason, William S.; Seeger, Christoph (Ed.):** Hepadnaviruses. Molecular Biology and Pathogenesis. 1991. 21 figs. X, 206 pp. ISBN 3-540-53060-6

Vol. 169: **Kolakofsky, Daniel (Ed.):** Bunyaviridae. 1991. 34 figs. X, 256 pp. ISBN 3-540-53061-4

Vol. 170: **Compans, Richard W. (Ed.):** Protein Traffic in Eukaryotic Cells. Selected Reviews. 1991. 14 figs. X, 186 pp. ISBN 3-540-53631-0

Vol. 171: **Kung, Hsing-Jien; Vogt, Peter K. (Eds.):** Retroviral Insertion and Oncogene Activation. 1991. 18 figs. X, 179 pp. ISBN 3-540-53857-7

Vol. 172: **Chesebro, Bruce W. (Ed.):** Transmissible Spongiform Encephalopathies. 1991. 48 figs. X, 288 pp. ISBN 3-540-53883-6

Vol. 173: **Pfeffer, Klaus; Heeg, Klaus; Wagner, Hermann; Riethmüller, Gert (Ed.):** Function and Specificity of γ/δ T Cells. 1991. 41 figs. XII, 296 pp. ISBN 3-540-53781-3

Vol. 174: **Fleischer, Bernhard; Sjögren, Hans Olov (Eds.):** Superantigens. 1991. 13 figs. IX, 137 pp. ISBN 3-540-54205-1

Vol. 175: **Aktories, Klaus (Ed.):** ADP-Ribosylating Toxins. 1992. 23 figs. IX, 148 pp. ISBN 3-540-54598-0

Vol. 176: **Holland, John J. (Ed.):** Genetic Diversity of RNA Viruses. 1992. 34 figs. IX, 226 pp. ISBN 3-540-54652-9

Vol. 177: **Müller-Sieburg, Christa; Torok-Storb, Beverly; Visser, Jan; Storb, Rainer (Eds.):** Hematopoietic Stem Cells. 1992. 18 figs. XIII, 143 pp. ISBN 3-540-54531-X